Urban Transportation Systems in Emerging Countries

This book offers the comprehensive analysis of urban transportation systems in emerging countries across Asia, South America, and Africa. It explores the characteristics, challenges, and opportunities of transportation modes; the integration of ICT innovations; modernization of operations; and the dynamic interplay between transportation and urban elements. This book provides practical approaches to building sustainable transportation systems tailored to the unique contexts of emerging countries.

Edited and authored by experts with extensive research and practical experience, this book is an essential resource for transportation planners, engineers, urbanist, policymakers, researchers, and development practitioners. It also serves as a foundational book for professionals and students aiming to address the complex challenges of urban transportation in emerging nations.

Through its insightful comparisons and analysis, this book enables readers to understand the challenges faced by transportation systems in emerging countries and derive lessons applicable to developed countries.

Mihoko Matsuyuki is Professor in Urban and Regional Planning at Yokohama National University. She has served as a member of Japan International Cooperation Agency's Advisory Committee on Environmental and Social Considerations.

Fumihiko Nakamura is Project Professor in the Graduate School of Frontier Sciences at the University of Tokyo. He is also a sub-program director in the Cross-ministerial Strategic Innovation Promotion Program organized by the Japanese Cabinet Office in the field of smart mobility platform development.

Urban Transportation Systems in Emerging Countries

Edited by
Mihoko Matsuyuki and Fumihiko Nakamura

Routledge
Taylor & Francis Group
LONDON AND NEW YORK

Designed cover image: Upper left photograph, Fumihiko Nakamura; lower left photograph, Mihoko Matsuyuki; upper right photograph, Shinji Tanaka; lower right photograph, Fumihiko Nakamura

First published 2026
by Routledge
4 Park Square, Milton Park, Abingdon, Oxon OX14 4RN

and by Routledge
605 Third Avenue, New York, NY 10158

Routledge is an imprint of the Taylor & Francis Group, an informa business

British Library Cataloguing-in-Publication Data
A catalogue record for this book is available from the British Library

ISBN: 978-1-032-72615-1 (hbk)
ISBN: 978-1-032-74044-7 (pbk)
ISBN: 978-1-003-46740-3 (ebk)

DOI: 10.1201/9781003467403

Typeset in Sabon
by Apex CoVantage, LLC

Contents

Notes on Editors xi
Notes on Contributors xiii
Preface xvii
MIHOKO MATSUYUKI AND FUMIHIKO NAKAMURA

1 Introduction 1
FUMIHIKO NAKAMURA AND MIHOKO MATSUYUKI

1.1 What Is Happening in Emerging Cities Now 1
1.2 Characteristics of Urban Transportation Systems in Emerging Countries 2
1.3 Expectations for Urban Transportation Systems in Emerging Countries 4
1.4 Cities and Transportation 5
1.5 Structure of This Book 6

PART I
Each Travel Mode and Its Space 11

2 Active Transportation and Its Space 13
2.1 From Cars to Active Transportation – Shaping Urban Spaces 13
MIHOKO MATSUYUKI

2.2 Street Vendors: An Obstacle or an Asset for Pedestrians? 16
SHINO MIURA

2.3 Redesign of Public Space and Pedestrianized Street, the Experience in Curitiba, Brazil 22
FUMIHIKO NAKAMURA

2.4 Shared Bicycle in China 25
SHENGCHUAN ZHAO

2.5 Conclusion 32
MIHOKO MATSUYUKI

3 Automobile, Motorcycle, and Freight 35

3.1 Transition from Bicycle and Motorcycle City to Car City 35
FUMIHIKO NAKAMURA

3.2 Motorcycle Safety Issue (Engineering, Education, and Enforcement) 38
FUMIHIKO NAKAMURA

3.3 Motorization and Road Development 40
TOSHIYUKI OKAMURA

3.4 Causes of Traffic Jam 44
SHINJI TANAKA

3.5 Traffic Regulation, Control, and Management 48
SHINJI TANAKA

3.6 Freight Transportation and Logistics 55
TAKAYOSHI FUTOSE

3.7 Conclusion 62
FUMIHIKO NAKAMURA

4 Paratransit in Emerging Countries 65
TOSHIYUKI OKAMURA

4.1 What Is Paratransit? 65
4.2 Variety of Vehicles 71
4.3 Operation and Regulations of Paratransit 73
4.4 Coordination and Integration of Paratransit with Institutional Mass Transit Systems 80
4.5 Future Perspectives on Paratransit 83

5 Bus and BRT 86

5.1 How Buses Are Used in Emerging Countries? 86
FUMIHIKO NAKAMURA

5.2 Introduction of BRT 88
FUMIHIKO NAKAMURA

5.3 BRT Management in Bogotá, Colombia 90
FUMIHIKO NAKAMURA AND YURIE TOYAMA

5.4 Reform of Bus Management System in Seoul, South Korea 96
FUMIHIKO NAKAMURA AND YURIE TOYAMA

5.5 Conclusion 101
FUMIHIKO NAKAMURA

6 Railway 104

6.1 Advantages of Urban Railways in Emerging Countries 104
TOSHIYUKI OKAMURA

6.2 Type of Urban Railway and Its Development 108
TOSHIYUKI OKAMURA AND FUMIHIKO NAKAMURA

6.3 Financial Management and Operating Body of Railway 116
TOSHIYUKI OKAMURA

6.4 Coordination of Railway Service with Other Transportation Modes 118
TOSHIYUKI OKAMURA

6.5 Conclusion 120
TOSHIYUKI OKAMURA

PART II
Transportation Management 123

7 New Services Aided by ICT 125

7.1 Overview of the Emergence of New Ideas 125
YURIE TOYAMA

7.2 CASE 127
YURIE TOYAMA

7.3 Ride-Hailing 135
MIHOKO MATSUYUKI

7.4 MaaS 138
YURIE TOYAMA

7.5 Conclusion 140
YURIE TOYAMA

8 Modernization of Passenger Transportation Management 144

8.1 Information Provision Integration 144
TOSHIYUKI OKAMURA

8.2 Network Improvement and Interchange Management 148
FUMIHIKO NAKAMURA

8.3 Ticketing Coordination 153
FUMIHIKO NAKAMURA

8.4 Financial Management 157
FUMIHIKO NAKAMURA

8.5 Conclusion 159
FUMIHIKO NAKAMURA

PART III
Urban Dimensions and Transportation Interplay: Planning, Equity, and Environment 163

9 Urban Transportation and Land Use 165

9.1 Relationship between Land Use and Transportation 165
MIHOKO MATSUYUKI

9.2 The Impact of Urban Railways on Preventing Urban Sprawl: Case of Bangkok, Thailand 170
MIHOKO MATSUYUKI

9.3 Evaluating the Effectiveness of Polycentric Urban Structure with Mixed-Use Development in Reducing Automobile Dependency: Case Study in Hanoi, Vietnam 175
MIHOKO MATSUYUKI

9.4 Transit-Oriented Development – Curitiba Case Study 182
FUMIHIKO NAKAMURA

9.5 Conclusion 185
MIHOKO MATSUYUKI

10 Urban Transportation and Equity 189

10.1 What Is Transportation Equity? 189
MIHOKO MATSUYUKI

10.2 Transportation Planning and Poverty Reduction in Curitiba 194
FUMIHIKO NAKAMURA

10.3 The Effects of Installing Cable Car System in Low-Income Area in Medellín 196
MIHOKO MATSUYUKI

10.4 Gentrification by Instalment of Urban Railway in Bangkok 201
MIHOKO MATSUYUKI

10.5 Conclusion 205
MIHOKO MATSUYUKI

11 Transportation Systems and the Environment 208

11.1 How Transportation and Environmental Issues Are Related? 208
FUMIHIKO NAKAMURA

11.2 Vehicle Electrification and Its Efficient Management 212
FUMIHIKO NAKAMURA

11.3 Demand Management to Reduce Transportation Sector Emissions 214
YURIE TOYAMA

11.4 Coordination of Spatial Planning and Mobility Design for Carbon-Neutral Society 216
FUMIHIKO NAKAMURA

11.5 Conclusion 218
FUMIHIKO NAKAMURA

Conclusion: Toward Sustainable and Inclusive Urban Transportation Systems in Emerging Countries 221
MIHOKO MATSUYUKI AND FUMIHIKO NAKAMURA

Index 225

Editors

Mihoko Matsuyuki is Professor in Urban and Regional Planning at Yokohama National University in Japan. Based on field surveys, she has researched cities in Asia and South America for more than 20 years. She has examined the integration of transportation and land use planning, including transit-oriented development (TOD) around rail and bus rapid transit (BRT) stations, the impact of rail development on urban structure and the pedestrian environment, and transportation and equity, including gentrification by rail and poverty reduction by transportation. In addition to her academic achievements, she has also contributed to urban planning practice in emerging countries, serving as a member for Japan International Cooperation Agency (JICA)'s Advisory Committee on Environmental and Social Considerations and an adviser on Davao Metropolitan Area Master Plan project in the Philippines. She also served urban planning practice in Japan as a member of the committees of many local governments such as the Tokyo Metropolitan Government's Urban Planning Committee. She is the editor of a book *Asian Urbanism in the Global Era* (in Japanese). She has extensive experience in publishing papers on urban planning and urban transportation in emerging countries: "Assessment of the sustainability of online food delivery from the perspective of CO_2 emissions and transport intensity: A case study in Jakarta" (2024) (with Fitrinitia, I., and Chotib, et al.); "Neoliberalization of urban planning and spatial inequalities in Asian megacities: Focus on Tokyo, Bangkok, Jakarta, and Mumbai" (2022) (with Kidokoro, T., and Shima, N.); "Impact of gentrification on travel behavior in transit-oriented development areas in Bangkok, Thailand" (2022) (with Aizu, N., Nakamura, F., and Leeruttanawisut, K.); and "Residents' travel behavior in a low-income settlement with aerial cable cars in Medellin, Colombia" (2022) (with Okami, S., Sarmiento-Ordosgoitia, I., and Nakamura, F.).

Fumihiko Nakamura is Project Professor of Urban Transportation Planning and Smart City Planning at Sustainable Society Design Center, Graduate School of Frontier Sciences in the University of Tokyo, Japan. He has worked as a professor in urban transportation planning and policies, some of which are related to issues in emerging countries. In his career, he was once a Visiting Professor at Graduate School of Urban Management, Pontifical Catholic University of Parana State in Brazil located in Curitiba, Brazil, in 2011; he was also once Assistant Professor at Human Settlement Development, Asian Institute of Technology, located in Thailand from 1992 to 1994. Currently, Professor Nakamura is a technical advisor in the field of smart city and smart mobility at JICA. He is author/editor of a number of books including *Peak-less Cities* (2023) (in Japanese), *Introduction to MaaS* (2022) (in Japanese); *Urban Transportation Planning: the 3rd edition* (2017) (in Japanese), *BRT Changes Cities* (2016) (in Japanese); and *Urban Transport and the Environment: International Perspective* (2004). He is also author/co-author of numerous articles published in international academic and professional journals, including "Spatial and time zone distribution of diverse pedestrian activities in the street network of Southeast Asian developing cities" (2023) (with Miura, S., Yoshida, M., Wongwiriya, P., and Tanaka, S.); "A study on the barriers of the Thai government for development of high speed rail project" (2019) (with Tissayakorn, K., Nakamura, F., Tanaka, S. and Miura, S.); and "Perspective on an urban transportation strategy with BRT for developing cities" (2017) (with Makimura, K. and Toyama, Y).

Contributors

Takayoshi Futose is an urban transport planner at Almec Corporation, a Japanese planning consultancy, holding a doctoral degree in Global and Regional Studies from Toyo University. He has been involved in urban transport master planning projects in emerging countries, including "the Comprehensive Study of the Urban Transport Development Program in Greater Yangon, Republic of the Union of Myanmar (JICA, 2019)"; "Data Collection Survey on Transportation Network for Regional Economic Development in Mekong Delta, the Socialist Republic of Vietnam (JICA, 2023)"; and "Empirical Research on the Social and Economic Impacts of Infrastructure Projects in the People's Republic of Bangladesh (JICA, 2023)." His publications include "Yangon's Urban Transportation: Traffic Demand and Its Quality of Service" (with M. Kojima, P. L. Binh, and H. Kato; 2015) and "Study on Private Vehicle-Dependence of Asian Developing Megacities Focusing on Mobility Gap between Vehicle Owners and Non-Owners" (with T. Okamura; 2022). He contributed specifically to Section 3.6 of this book.

Shino Miura is Doctor of Environmental Studies and an associate professor in the Department of Civil and Environmental Engineering, Faculty of Science and Engineering of Chuo University, Japan. She has worked as a researcher on livable city policy and its diffusion. Her work included committee members of Japanese national guidelines on street design (Ministry of Land, Infrastructure, Transport and Tourism) as well as pilot projects in Asian cities. The first project was a student workshop and tactical action in Khon Kaen City with the local university in Thailand. Currently, she is responsible for advising a JICA project on the main street placemaking in Luang Prabang, Laos. Her recent publications include Development of Workshop Framework Empowering Local Stakeholders for "Place Strategy" in a District: An Implementation at Omotesando, Tokyo (co-authored with H. Usui, Y. Ishida, K. Yamada, and Y. Oyabu, Research in Transportation Economics, Thredbo 17 Conference – Special Issue, 2023). She contributed specifically to Section 2.2 of this book.

Toshiyuki Okamura is a professor in the Faculty of Global and Regional Studies at Toyo University, Tokyo, Japan. His research interests are urban transport planning, public transportation planning, and operation from global and local perspectives. He has conducted academic surveys in many emerging countries such as Cambodia, China, Fiji, Kenya, Pakistan, the Philippines, South Africa, Sri Lanka, Thailand, and Vietnam. He is an author of many journal papers on transportation planning issues in the emerging world, including "Analyzing behavioral intentions in new residential developments of motorcycle dependent cities: The case of Ho Chi Minh City, Vietnam" (with Le Quan HOANG; 2020); "Transit Oriented Development Precincts: A Study of Various Potential TOD Precincts Surrounding Gautrain Stations in Johannesburg South Africa" (with MALOBA Tshegofatso; 2019); and "Factors Influencing the Acceptability of Travel Demand Management Measures in Lahore: Application of Behavioral Theories" (with Muhammad Ashraf JAVID; 2015). He contributed specifically to Section 3.3, Chapter 4, Chapter 6, and Section 8.1 of this book.

Shinji Tanaka is a professor in the field of Traffic Engineering at Yokohama National University (YNU), Japan. He received Doctor of Engineering from the University of Tokyo (UT), Japan, and has worked for that university before joining YNU. His research interests are traffic operation and management, traffic modeling and simulation, Intelligent Transport Systems (ITS), and so on. He has experience of serving as a Visiting Assistant Professor in Asian Institute of Technology (AIT), Thailand, and conducting a lot of studies in many emerging countries in Asia, Africa, and South America, such as mixed traffic flow analysis, traffic accident analysis, on-street parking management, and signalized intersection control. He published many academic papers as well as books including *Spatial Data Infrastructure for Urban Regeneration* (with Y. Sadahiro; 2008) and *International Encyclopedia of Transportation* (with R. Vickerman; 2021). He contributed specifically to Sections 3.4 and 3.5 of this book.

Yurie Toyama is a designated assistant professor in the Institutes of Innovation for Future Society, Global Research Institute for Mobility in Society (GREMO) at Nagoya University, Japan, and also Project Research Associate, Institute of Industrial Science, University of Tokyo. Her research interests include urban transportation planning, public transport improvement especially for bus systems with new technology including automated driving, and Mobility as a Service (MaaS). Her research field extends from Japan to Southeast Asian countries and South American countries such as Vientiane, Jakarta, Bangkok, Bogota, and Curitiba. She has worked as a member of the research project "A study on directions of urban policies and road transport policies in Japan based on development trend on smart cities and CASE related technologies in Asian region," funded by

The Japan Research Center for Transport Policy. Her publications include "A Study on BRT Applicability on Large Cities in Developing Countries" (2014; with Fumihiko Nakamura). She contributed specifically to Sections 5.3, 5.4, 7.1, 7.2, 7.4, 7.5, and 11.3 of this book.

Shengchuan Zhao is Professor of Graduate School of Business Design at Josai International University, Japan. He was Research Fellow of Japan Society for the Promotion of Science, Fulbright Scholar at Harvard Kennedy School and Professor at the Dalian University of Technology, China. He is the recipient of International Outstanding Collaboration Award of Japan Society of Civil Engineers (JSCE) and an overseas fellow of International Association of Traffic and Safety Sciences (IATSS). He was elected a Foreign Fellow of the Engineering Academy of Japan (EAJ) in 2025. His research interests are urban transportation planning, urban transportation economics, and transportation issues for emerging countries. Professor Zhao has published numerous papers in refereed journals including "Understanding and modeling willingness-to-pay for public policies to enhance road safety: A perspective from Pakistan" (with Subhan, F., Ali, Y., and Oviedo-Trespalacios O.) (2023) and "The evolution of spatial equity of high-speed rail accessibility in China: An operation frequency based approach" (with Luo, H., Cui, M., Zhong, S., and Ma, J.; 2022). He contributed specifically to Section 2.4 of this book.

Preface

Mihoko Matsuyuki and Fumihiko Nakamura

Motivation, Expected Readers, and Aims of This Book

The editors and contributors of this book are experts in transportation and urban planning who have conducted research and practical projects in emerging countries across Asia, South America, and Africa. Through these activities, we have come to deeply understand the severity of urban transportation challenges in these regions. Stepping into a city in an emerging country, we are immediately struck by the severity of air pollution and overwhelmed by the crippling traffic congestion and frequent traffic accident. Within this environment, people spend long hours navigating through traffic, suffer health problems caused by air pollution, and face limited mobility options that restrict access to education and employment opportunities. These issues highlight how urban transportation problems not only impede national economic development but also greatly affect individual well-being.

While investigating urban transportation problems in emerging cities, we realized that these challenges are uniquely shaped by the context of emerging countries. They cannot be fully explained by theories and models developed in developed countries. Although the urban transportation systems of emerging countries may appear less developed or inferior compared to those in developed countries, they often offer innovative and advantageous solutions, such as various transportation services collectively referred to as "paratransit." These services provide flexible and affordable almost door-to-door options that are uncommon in developed countries. This realization has driven us to explore the underlying factors behind urban transportation issues in emerging countries and the fundamental relationship between cities and transportation.

Based on these experiences, we strongly believe that professionals involved in development and management of urban transportation systems in emerging countries – planners, engineers, policymakers, researchers, as well as donors – must go beyond transportation theories and models developed in developed countries. They need to understand the unique problem structure of emerging countries to effectively plan their urban transportation systems.

However, there is a significant lack of books addressing urban transportation systems in emerging countries, particularly recent publications that reflect the rapidly evolving urban environments and technological advancements in these regions. This gap motivated us to write this book.

This book aims to provide insights into urban transportation systems in emerging countries by analyzing the characteristics, challenges, and contexts of different travel modes; the adaptation of new technologies on transportation systems; modernization of operations; and interplay between transportation and urban dimensions. It also offers practical approaches to building sustainable e transportation systems from the perspective of emerging countries. The goal is to provide readers with a comprehensive understanding and an entry point for deeper exploration of urban transportation systems in emerging countries.

The primary reader for this book includes professionals such as planners, urbanist, engineers, administrators, and researchers, involved in urban transportation in emerging countries. These individuals may be local experts or professionals from developed countries acting as donors or advisors. The demand for professionals in this field far exceeds the current supply. We hope this book will serve as a textbook for current and future professionals in urban transportation for emerging countries. Moreover, by addressing the relationship between transportation and various urban dynamics in emerging countries, the book provides insights not only for transportation professionals but also for those in urban planning and related fields.

Even for students and practitioners currently uninterested in urban transportation systems in emerging countries, this book offers a valuable opportunity to understand these systems comprehensively and compare the challenges of urban transportation in emerging and developed countries. Such comprehensive understanding and comparisons provide objective perspectives on urban transportation systems in developed countries and lessons from the emerging countries.

While the urban transportation needs of emerging countries are vast and diverse, and the challenges numerous, research interest in this field remains relatively low. Through this book, we hope to inspire a broader interest in the urban transportation systems of emerging countries and foster diverse perspectives. Our aspiration is that this book serves as a stepping stone for readers to deepen their engagement with these topics, leading to meaningful learning and practical applications. Ultimately, we hope this book contributes to the establishment of sustainable urban transportation systems in emerging countries and, through these efforts, enhances the well-being of the people living in these cities.

The insights presented in this book should be further developed through practical application in the field. Readers are encouraged to use this book as a reference for various initiatives and to share the outcomes as feedback. Such feedback will serve as an indispensable resource for the continued evolution of urban transportation planning. In future editions, we aim to incorporate these insights, reflecting the feedback to contribute to further advancements in urban transportation systems. It is our hope that this book will serve as a stepping stone toward achieving sustainable urban transportation.

It is worth noting that this book uses the term "emerging countries" rather than "developing countries." By "emerging countries," we refer to nations whose economic development is approaching that of developed countries and which face challenges shared by both developing and developed countries. However, the insights presented in this book are broadly applicable to developing countries as well.

Chapter 1

Introduction

Fumihiko Nakamura and Mihoko Matsuyuki

1.1 WHAT IS HAPPENING IN EMERGING CITIES NOW

The landscapes of large cities in emerging countries are undergoing dramatic transformations. This rapid change is undoubtedly driven by the rapid economic development occurring in these nations. Urban areas, particularly megacities, serve as the engines of economic growth in emerging economies. These cities attract a large number of people migrating from smaller cities, from rural areas, and sometimes even from abroad, in search of better employment opportunities. Consequently, urban populations are surging, and cities are expanding significantly. However, this rapid urbanization brings severe challenges, many of which are deeply intertwined with transportation systems.

According to the TomTom Traffic Index 2024, Mexico City, Mexico, ranks as the city with the highest congestion level globally. It is followed by cities in emerging countries, such as Bangkok (Thailand), Davao City (Philippines), Bandung (Indonesia), and Pune (India) (TomTom International BV). Traffic congestion is a particularly severe issue in the megacities of emerging countries, resulting in significant economic losses and profoundly affecting the quality of life for citizens.

The causes of such severe congestion include delays in road construction to meet growing populations; underdeveloped mass transit systems like railways; the rapid progression of motorization; and inadequate traffic management systems such as signaling, inefficient road networks, and poor road maintenance. These factors collectively contribute to the critical state of traffic congestion in emerging cities.

In addition to severe traffic congestion, air pollution caused by vehicle emissions has become a critical issue, exacerbated by poorly maintained vehicles, the use of substandard fuel, and the ineffectiveness of emission regulations. Cities in emerging countries, such as those in India and China, dominate the rankings of the most polluted cities worldwide (AQI India). The health impacts of this air pollution are severe, contributing to respiratory and cardiovascular diseases (Mannucci and Franchini, 2017). It is estimated

DOI: 10.1201/9781003467403-1

that air pollution causes approximately 2 million premature deaths annually, highlighting the urgent need to address this pressing issue.

Moreover, in emerging countries with significant economic disparities and large populations living in poverty, the lack of well-developed mass transit systems exacerbates the issue. While affluent individuals rely on private cars for transportation, the poor are often forced to use buses or paratransit services, requiring multiple transfers and prolonged travel times. Some low-income individuals lack access to any means of transportation, leaving them unable to travel. This limits their opportunities for employment, education, and health care, further entrenching poverty and widening economic inequality.

Thus, transportation issues in emerging cities are even more severe than in developed countries, with far-reaching and diverse impacts. Improving transportation systems in these countries is an urgent priority.

1.2 CHARACTERISTICS OF URBAN TRANSPORTATION SYSTEMS IN EMERGING COUNTRIES

The transportation issues in the cities of emerging countries, which contribute to such severe urban problems, differ significantly in their background and structure compared to those in developed countries. Therefore, it is crucial to understand the unique context and problem structure when formulating transportation policies and plans.

The progression of motorization can be attributed to several fundamental factors, such as the expansion of road network, the lack of mass transit, the emergence of a growing middle class able to afford cars due to economic development, and urban sprawl – factors similar to those seen in developed countries. However, unique to emerging countries is the tendency for car ownership and use to symbolize high social status. As a result, many people in these countries are willing to take on heavy loans to own cars, further accelerating motorization. Conversely, the use of public transportation is often perceived as a marker of lower social status, sometimes leading to outright discrimination based on social hierarchy. In some nations, cultural or religious contexts also influence these dynamics. To dislodge such entrenched perceptions, public transportation vehicles, stations, and services require tailored approaches distinct from those in developed countries. Additionally, safety concerns further encourage car use, with child drop-offs and pick-ups at schools by car contributing significantly to traffic congestion during morning and evening peak hours in many cities.

The timing of motorization and railway introduction has also had a significant impact. In large cities in Europe, the United States, and Japan, urban railways were introduced during the early stages of urbanization, and urban areas expanded alongside the growth of railway networks. Motorization followed later, after railways had already been established. By contrast, in emerging countries, urbanization has progressed at an exceptionally rapid pace, but due to geographical and historical circumstances, modernization has lagged behind. Consequently, railway systems were introduced much later, allowing motorization to advance first. This led to urban sprawl along major roads, resulting in transportation systems and land use patterns that are heavily car-dependent. When railways are eventually introduced, the pre-existing urban sprawl makes it challenging to lay tracks on flat structures or secure station plazas and transfer spaces around stations. This hampers the convenience of railway use, making modal shifts from cars to railways more difficult to achieve.

Another significant difference between the transportation systems in emerging and developed countries lies in the dual economic structure. In emerging economies, both the formal and informal sectors coexist and play crucial roles in supporting the economy. In the transportation sector, paratransit, once categorized as informal transportation, plays a major role. Due to its affordability, paratransit is an indispensable means of mobility for low-income populations. It supplements formal public transportation by serving areas and times that formal systems cannot adequately cover. However, paratransit faces significant challenges in terms of safety and reliability. Additionally, many paratransit services operate without official registration, making it difficult to incorporate them into government transportation plans and regulations. Often run by individuals or small-scale operators, these services are also characterized by difficulties in data collection and management.

One notable characteristic of urban and transportation planning in emerging countries is the rapid pace of societal changes, leading to frequently shifting prerequisite and high levels of uncertainty in planning. Political instability in some countries often prevents transportation infrastructure projects from being executed as planned due to regime changes, coups, or civil conflicts. For instance, the Japanese government's Overseas Development Aid (ODA)-supported project to upgrade Myanmar's core railway network was significantly scaled back following the 2021 coup and transition to a military regime, which halted additional ODA funding. Emerging countries are also highly sensitive to external factors, such as fluctuations in exchange rates and global economic instability, partly because they often rely on funding from ODA or foreign companies through public-private partnerships (PPP). Economic turbulence can quickly lead to delays, interruptions, or reductions in infrastructure projects, underscoring the vulnerability of planning to external changes. Additionally, governance issues, such as insufficient governmental capacity and corruption, further complicate planning reliability. Changes in stakeholder values also add to the uncertainty. For example, policies prioritizing the preservation of historical assets or promoting environmentally friendly actions could significantly alter decisions on route alignments and construction procedures for road and rail projects. These are challenges that developed countries have also faced over the past few decades. Similarly, attitudes toward private vehicle ownership might evolve in the future, highlighting the importance of considering potential shifts in stakeholder values.

In many emerging countries, transportation-related legal frameworks, such as road traffic laws and air pollution control regulations, remain underdeveloped. Even where laws exist, enforcement systems are often inadequate, leading to ineffective implementation of driver licensing systems, traffic violation penalties, emissions controls, and vehicle age restrictions. Simply establishing regulations is insufficient; attention must also be given to enforcement and management systems. Furthermore, land-use regulations in these countries are often looser than those in developed countries or, even when rigid on paper, are poorly enforced. This can result in inefficient land use and the proliferation of urban sprawl. Addressing these regulatory and enforcement challenges is critical for sustainable transportation systems in emerging countries.

The circumstances in emerging cities differ significantly from those in developed countries. Therefore, applying transportation theories, models, or successful practices derived from developed countries directly to emerging nations is unlikely to yield the same outcomes. It is essential to first understand the background and structure of problems specific to emerging countries before formulating and implementing appropriate solutions.

1.3 EXPECTATIONS FOR URBAN TRANSPORTATION SYSTEMS IN EMERGING COUNTRIES

The transportation issues in emerging cities are exceptionally severe, with underlying dynamics distinct from those in developed countries. As a result, the direct application of theories or strategies designed for developed cities often proves ineffective. Addressing transportation problems in cities in emerging countries is inherently more complex and challenging than in their developed counterparts.

However, the rapid pace of change and the relative looseness of regulations in emerging countries can also be viewed as opportunities. The lack of established infrastructure and systems provides a certain flexibility, creating room for innovative initiatives. Furthermore, the strong authority often held by governments in emerging countries enables bold and experimental policy implementations.

This flexibility has facilitated social experiments that can sometimes lead to significant advancements. For example, in Bogotá, Colombia, a program called Ciclovía transforms major city roads into car-free spaces for bicycles, running, and walking every Sunday and holiday from 7 a.m. to 2 p.m., spanning approximately 120 kilometers. Initially introduced in 1974 as a small-scale experimental initiative, it has since evolved into a permanent program across Bogotá. Today, similar programs inspired by Ciclovía have been adopted in other cities, including New York (Summer Streets), Mexico City (Muévete en Bici), and São Paulo.

In cities in emerging countries, infrastructure development tends to lag behind. However, unlike developed countries, where infrastructure was gradually built and upgraded alongside technological advancements, emerging cities often experience a phenomenon known as leapfrogging, whereby the latest technologies are adopted rapidly and on a large scale. For instance, in developed countries, telephone networks were initially established with landlines, which gradually evolved into mobile phones and eventually smartphones. In Bangkok, the installation of landline networks was delayed, and before landlines became widespread, mobile phones were already widely adopted in the 1990s, ahead of many developed nations. Similarly, in modern-day Africa, smartphones have become prevalent without the widespread adoption of landlines.

The rapid penetration of information and communication technology (ICT), particularly smartphones, has enabled the swift introduction of innovative services, some of which have experienced explosive growth in certain cities. For example, in China, bicycle-sharing services quickly gained widespread popularity. In Southeast Asia, particularly in countries like Indonesia, ride-hailing (RH) services have expanded rapidly and become indispensable for urban mobility. Additionally, these technologies have evolved uniquely by integrating with paratransit systems, creating tailored solutions for local contexts. However, while new technologies can be adopted quickly, they also bring significant challenges. For instance, in China, the large-scale disposal of unused shared bicycles has become a pressing social issue. Developed countries, which tend to implement new technologies more cautiously, have much to learn from the rapid adoption and subsequent developments in countries and cities that have been quick to embrace and scale these technologies. This book delves into the adoption, development, and challenges of new technologies applied to transportation system that have been rapidly embraced and advanced in emerging economies.

1.4 CITIES AND TRANSPORTATION

Cities are dynamic spaces where social, economic, and environmental elements are intricately intertwined. Urban transportation does not operate independently; instead, it functions in close connection with various elements in cities. To understand transportation phenomena, it is crucial to analyze how transportation systems work within the broader urban framework and how they influence and are influenced by other elements. For example, severe traffic congestion is not merely an issue of traffic volume. Its underlying causes often involve land use patterns, the development of public transportation infrastructure, societal values, and complex socio-economic contexts. Without understanding these factors, effective solutions cannot be developed.

Moreover, cities are constantly undergoing significant changes. The future of urban transportation should not be discussed in isolation but rather considered within the broader context of how cities should be shaped moving forward. In emerging countries, urban changes are particularly significant, with rapid transformation and highly complex dynamics within cities. Through this book, we aim to explore the dynamic and intricate relationships between urban systems and transportation in emerging countries and examine pathways toward sustainable cities and urban transportation systems.

1.5 STRUCTURE OF THIS BOOK

The book is divided into three main parts. Part I focuses on the various modes of urban transportation and their associated spaces. Part II discusses the technologies and management of urban transportation systems. Part III explores the interactions between different urban dimensions and transportation. Part I comprises five chapters, Part II contains two chapters, and Part III includes three chapters, along with this chapter and the concluding chapter, making up a total of 12 chapters.

Chapter 2 discusses about active transportation and its space. It introduces walking and cycling as key elements of active transportation. It focuses on the roles and uses of walking and cycling, as well as the social, economic, and environmental significance of pedestrian spaces. Using examples from Southeast Asia, the chapter discusses the relationship between the "place" functions of pedestrian spaces and their "space" functions, along with the challenges these spaces face. Additionally, it examines the redesign of public spaces and pedestrian areas in Curitiba, exploring their significance and associated issues, as well as the development, challenges, and proposed solutions for bicycle-sharing systems in China. Through these discussions, the chapter considers the future of active transportation in emerging countries.

Chapter 3 provides a comprehensive overview of motorized transportation, motorcycle and car, their space road, and their management and logistics. The first section explores the historical shift from bicycle- and motorcycle-based cities to car-dominated urban environments and its implications for urban infrastructure. The second section discusses traffic safety issues, highlighting the impact of motorization on accident rates and safety measures in emerging countries. The third section addresses issues in road development, including inadequate infrastructure and safety concerns, as motorization progresses. The fourth section analyzes the causes of traffic congestion and strategies for managing demand and optimizing road use. The fifth section examines transportation regulations, enforcement, and management approaches aimed at addressing urban mobility challenges. The final section explores urban freight transportation, highlighting the role of technology and logistics strategies in improving efficiency and sustainability.

Chapter 4 focuses on paratransit. Paratransit is a public transportation mode that provides important mobility in emerging countries, although it is less organized with relatively small vehicles and has many "problems." The first section defines paratransit, a mode particular to emerging countries, by outlining its technical and institutional position as an urban transportation mode. The second section presents the status of various paratransit modes, focusing on the global diversity and universality of paratransit. The third section elaborates on the operational characteristics of paratransit, focusing mainly on the relationships and informality among the various stakeholders. The fourth section describes the coordination and integration of paratransit with institutional mass transit systems and the formalization of paratransit. The fifth section describes future perspectives on paratransit, including the rise of ride-hailing and the potential of new technologies.

Chapter 5 explores the roles and management of buses and Bus Rapid Transit (BRT) systems in urban transportation, with case studies from various cities. The first section discusses urban buses, focusing on their roles in public transportation, particularly in emerging countries, and on examining the diversity of vehicle types and operations. The second section introduces the concept of BRT, detailing its advantages over traditional bus systems and its global adoption as an efficient urban transit solution. The third section examines the innovative BRT system in Bogota, Colombia, analyzing its infrastructure, operations, and impact on urban mobility. The fourth section explores Seoul's bus system reform, emphasizing its integration of ICT, advanced fare systems, and urban transportation policies.

Chapter 6 examines urban rail systems in emerging countries as a vital solution to urban transportation challenges, focusing on their development, integration, and operational strategies. The first section introduces the significance of urban rail in addressing congestion and environmental issues in densely populated cities. The second section explores the types of urban rail systems, including metros, light rail, and commuter trains, and their applications in different urban contexts. The third section discusses the integration of rail systems with other transportation modes to create seamless, multimodal networks. The final section highlights financial and operational challenges, emphasizing the importance of innovative funding models and effective management for sustainable development.

Chapter 7 examines the transformative impact of ICT on urban transportation, with a focus on emerging countries. It begins by introducing the rapid adoption of smartphones in Southeast Asia and the role of smart city initiatives in facilitating ICT-driven changes in mobility. The chapter then delves into the CASE framework (Connected, Autonomous, Shared, and Electric), which represents key innovations in the automotive industry and their applications in urban transportation. Next, it examines ride-hailing and ride-sharing services, emphasizing their evolution, challenges, and integration into urban mobility systems in SEA. Finally, the potential of Mobility as a Service (MaaS) is discussed, highlighting its ability to integrate diverse transportation modes, enhance user convenience, and promote sustainable transportation systems. Through these discussions, the chapter sheds light on how ICT and emerging technologies are reshaping transportation landscapes in emerging countries.

Chapter 8 investigates the modernization of passenger transportation management by addressing key aspects such as information provision, network planning, fare coordination, and financial strategies. The first section discusses the integration of transportation information systems, highlighting the importance of digitizing data and improving access to static and dynamic information, particularly in emerging economies. The second section examines network improvement and interchange management, emphasizing the role of master planning in creating effective transportation systems and ensuring connectivity. The third section focuses on fare coordination, including challenges related to unified fare systems, the introduction of smart payment methods, and the development of MaaS. Finally, the chapter analyzes financial strategies for sustainable transportation, including international aid, environmental policy frameworks, and revenue from special taxes, providing insights into funding mechanisms for urban transportation development in emerging countries.

Chapter 9 focuses on the interplay between urban transportation planning and land use, emphasizing their interdependence in shaping sustainable urban development. Section 9.1 introduces the basic concepts linking land use and transportation, discussing land use strategies to reduce car dependency. Section 9.2 examines the impact of urban railways on residential development patterns in Bangkok, highlighting its role in curbing urban sprawl. Section 9.3 explores Hanoi's polycentric urban structure and mixed-use developments, analyzing their influence on travel behavior and motorization. Section 9.4 reviews Curitiba's practices in transit-oriented development (TOD), focusing on the effects and challenges of integrating public transportation and land use policies.

Chapter 10 explores equity in urban transportation. It begins by introducing the concepts of horizontal equity, which emphasizes equal distribution of benefit and burdens of transportation among different groups, and vertical equity, which focuses on providing additional support to disadvantaged groups. Next, it reviews Curitiba's innovative bus system, outlining its efforts to support low-income groups and its challenges. The chapter then examines Medellín's Metrocable system, investigating its impacts on accessibility for low-income residents, commuting, economic opportunities, and pride for their neighborhood. Finally, it analyzes whether gentrification has occurred along Bangkok's urban rail lines and how travel behavior has changed among new and long-term residents.

Chapter 11 reviews the intricate relationship between transportation systems and environmental concerns, focusing on strategies for reducing environmental impacts. The first section explores the broad concept of "environment" and its various interpretations, particularly in relation to transportation's effects on air quality, noise, and greenhouse gas emissions. The second section introduces vehicle electrification as a key trend, discussing the adoption of electric buses and electric three-wheelers and their implications for sustainability. The third section delves into transportation demand management (TDM) strategies, including reducing travel demand and encouraging modal shifts to environmentally friendly options. The final section discusses the coordination of spatial planning and mobility design to achieve a carbon-neutral society, highlighting trade-offs and practical solutions for integrating environmental and transportation goals.

The Conclusion chapter summarizes the discussions from the preceding chapters. It then integrates these insights to discuss future directions for establishing sustainable transportation systems in emerging countries.

REFERENCES

AQI India. (n.d.). *Most polluted cities in the world*. Retrieved January 4, 2025, from https://www.aqi.in/world-most-polluted-cities

Mannucci, P. M., & Franchini, M. (2017). Health effects of ambient air pollution in developing countries. *International Journal of Environmental Research and Public Health*, *14*(9), 1048. doi:10.3390/ijerph14091048

TomTom International BV. (n.d.). Traffic index ranking. *TomTom*. Retrieved January 4, 2025, from https://www.tomtom.com/traffic-index/ranking/

Part I

Each Travel Mode and Its Space

Chapter 2

Active Transportation and Its Space

2.1 FROM CARS TO ACTIVE TRANSPORTATION – SHAPING URBAN SPACES

Mihoko Matsuyuki

2.1.1 Shifting from Cars to Active Transportation and Re-evaluation of Value of Pedestrian Space

At present, in many developed countries, particularly in Europe, the value of active transportation modes such as walking and cycling is being re-evaluated. There is a growing movement in numerous cities to create urban environments that support walking and cycling, aiming to enhance urban well-being, improve sustainability, reduce environmental impact, and promote social inclusion (Lah, 2019). To achieve this, many cities are reallocating road space; reducing areas designated for automobiles; and transforming them into spaces for pedestrians, cyclists, and public plazas. In this way, the trend in cities around the world is aiming to shift from cars to active transportation. For example, in Barcelona, multiple blocks are being reimagined as one large block (i.e., a "superblock"), and the speed of cars within the superblock is being restricted to low speeds, with the roads and intersections within it being used as living spaces by citizens. In this way, the value of active transportation and its space are currently being re-evaluated around the world.

2.1.2 Role of Pedestrian Space

Pedestrian spaces have various roles, not just for people to walk in. The following are the main roles (Monheim, 2003; McMillen, 2003; National Association of City Transportation Officials, 2016; Lah, 2019):

1. Enhancing Accessibility and Mobility

   Pedestrian streets provide better accessibility and mobility for people on foot. By restricting or prohibiting motor vehicle traffic, these streets

DOI: 10.1201/9781003467403-3

create a safer and more comfortable environment for walking, which can encourage more people to walk rather than drive.

2. Supporting Local Businesses

   Pedestrian streets often enhance the volume of shopping and other business activities in the area. With increased foot traffic, local businesses can see a boost in customers, which can lead to economic growth and revitalization of urban areas.

3. Promoting Social Interaction and Community Well-being

   These streets offer spaces for social interaction, allowing people to meet, socialize, and engage in community activities. This can foster a sense of community and improve overall well-being.

4. Providing Recreational Opportunities

   Pedestrian streets can be used for various recreational activities such as promenading, dining, performing, and other leisure activities. They create vibrant public spaces where people can relax and enjoy their surroundings.

5. Improving Environmental Quality

   By reducing vehicle traffic, pedestrian streets can lower noise and air pollution levels, contributing to a cleaner and more pleasant urban environment. This can also make the area more attractive to both residents and tourists. Furthermore, the pedestrian space provides a place for street trees, and also contributes to the greening of the city.

6. Enhancing Safety

   Pedestrian streets reduce the risk of accidents involving vehicles and pedestrians, making urban areas safer for all users, including children and the elderly.

In this way, pedestrian spaces play various roles in order for people to live safely, comfortably and sustainably in cities.

2.1.3 Pedestrian Space in Emerging Countries

In emerging countries, as the share of walking trips is higher than that in developed countries, the importance of pedestrian spaces is high. However, in many emerging countries, pedestrian spaces are not well provided for, or even if they are provided for, they are not well maintained, and there are often obstacles that make the environment insufficient for pedestrians (Figure 2.1). Furthermore, in areas where flooding is frequent due to the lack of sewerage systems, sidewalks are often raised significantly above the road level to prevent flooding of buildings, making it difficult for wheelchair users and people with walking difficulties to travel around.

Figure 2.1 Obstacles and potholes on the sidewalks of Bangkok

Source: (Photos taken by Nanami Aizu)

In this way, the environment for pedestrians is not well developed in emerging countries. Nevertheless, in many cities in emerging countries, the pedestrian streets are not only spaces for walking, but are also widely used as places for commerce (e.g., shops and food stalls), eating, recreation, and communication, and are extremely lively places (Figure 2.2).

However, in some cities in emerging countries, street food stalls have been banned due to concerns about hygiene, city beauty, and obstruction to pedestrians. For example, in Bangkok, street food stalls were banned on all major roads in the city centre in 2017 for the sake of public order and hygiene. Furthermore, in emerging countries, as motorization is progressing rapidly, pedestrian space is being taken away to make way for space for cars. This trend seems to be running counter to the global trend of promoting active transportation. So, how can we re-recognize the value of pedestrian space in emerging countries and maintain the quantity and quality of pedestrian space?

In this chapter, based on the aforementioned problem awareness, we will introduce the case of Khon Kaen in Thailand and Da Nang in Vietnam, which show the relationship between food stalls and walkability, the case of Curitiba in Brazil, which converted the carriageway into pedestrian space. In addition, we will discuss the use of bicycles as another form of active transportation, and look at how bicycle use has changed in China.

Figure 2.2 Activities on a sidewalk in Hanoi

Source: (Photos taken by Mihoko Matsuyuki)

2.2 STREET VENDORS: AN OBSTACLE OR AN ASSET FOR PEDESTRIANS?

Shino Miura

2.2.1 Background

Streets are responsible for transportation but also function as "places" that represent local livelihood culture. In Southeast Asian cities, many such places are generated by mobile vendor stalls and living spaces. The recent fundamental evaluation of the "place function" is polarized into negative and positive aspects. For instance, in Bangkok, Thailand, an administrative policy in the latter half of the 2010s sought to move food vendors off the streets to limit profitable private uses of public space. However, the continuity of vendor business on the streets was strongly supported because it enabled shopkeepers, including young people and women, to make a living and create cultural elements of the city. The competition for available space between people and vehicles or between places to linger and spaces for through traffic (e.g., walking) is intensifying in cities where the inclusive place function of public streets is required to support the growth of local livelihoods.

The author presumes that one of the factors for the emergence towards the exclusion is that the actual activities and spatial boundaries of local places where many different types of activities are mixed, that is, highly diverse, have not been clearly observed. When the content and quality of a target are unclear, management costs and risks increase, which leads to motivation of

Figure 2.3 Publicly maintained bench (left) and privately utilized sidewalk (right) in Khon Kaen city

Source: (Photos taken by Shino Miura)

whole clearance. In Southeast Asian or East Asian countries, even religious places with trees and statues watching over people have blended in quietly with the street corners. The built environment has been formed that allows daily activities to seep out as if they were the natural course of things, such as shophouse-style houses and cul-de-sacs and alleys like the "soi" in Thailand. Children play there and develop social skills. While street design and traffic engineering are specialist fields that have been introduced and developed in the West, we should listen to the subjective opinions of local people when it comes to thinking about such a compositive quality of places.

For instance, here is the result of a survey we previously conducted among local university students in Khon Kaen, Thailand, which is one of the main targets of this chapter: We showed the answerers photos of publicly maintained bench area (at bus stop) and sidewalks with a concentration of street vendors and asked them for their subjective impressions (see Figure 2.3). Both need improvement in maintenance, the marketized street, which is larger in terms of the number of goods occupying the street space, received better ratings in indicators such as ease of "preferable to stay", "connected to the neighbourhood", and "active" scenery. Over 70% of these place evaluation (answerers) judged by the appearance of people/business using it (see Figure 2.4). This suggests that we should not only judge the place by the quantity of things occupied but also look at people's on-site activities.

2.2.2 Research Objective and Target Cities

With this experience in mind, we focused the on-site activities in street places in two Southeast Asian cities. The standards of street design should sensitively control spaces, meet the diverse demands of citizens, and examine possible redistributions of spaces instead of making a choice between accepting and removing places. The research objective was to clarify citizens' demand

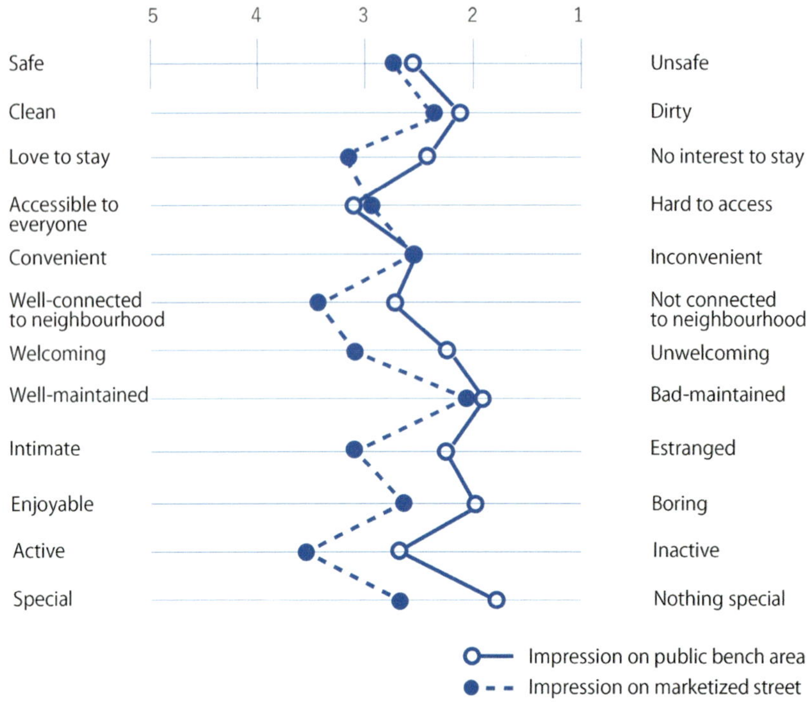

Figure 2.4 Average scores of the impression on public bench area and marketized street by local university students (n = 50, 96% of them were daily users)

of places and the spatial distribution of places in the Central Business District (CBD). After considering local academic organizations' support for the research, we selected two target cities: Da Nang (1.3 million population), a coastal city in Vietnam with a grid-like network whose history includes French colonialism; and Khon Kaen in Thailand (1.8 million population), which was originally an assembly of villages adjacent to a lake and grew after the railway was constructed. Da Nang has applied a policy of better traffic control and sidewalk extension, while Khon Kaen employs student guides for peak-hour pedestrians and operates a temporary car-free zone for the night market.

2.2.3 Hypothesis and Research Methods

In the first place, is there a conflict between pedestrians and stationary activities including vendor stalls and their users?

For instance, the off-peak hour of congestion index (ppmm value = people per hour ÷ 60 ÷ clear footway width [m]) suggests that the sidewalk of

Main Street in Khon Kaen, where diverse stalls exist, is sufficiently comfortable. This index is used by an advanced transportation street policy-maker, Transport for London, in their guideline on the pedestrian environment. A calculated ppmm of less than 8.0 means the street is comfortable because less than 30% of it is restricted for walk-through movement. It will be crucial to evaluate the situation based on the timeframe, such as whether walking or movement is hindered during peak commuting hours (around 16:30–18:00). Therefore, we conducted on-site research to clarify the spatial and timeframe distribution of pedestrian activities in the street network of CBD areas.

This study involved conducting a literature review, on-site investigation, and interviews about street management in the target cities. Three hypotheses regarding the distribution of pedestrian stationary activities were established from the results:

Hypothesis 1: Timeframe when stationary activity is extensive
Stationary activity, especially optional activity, might be high during times other than daytime when the temperature is high (including peak commuting hours).

Hypothesis 2: Segments with a large amount of stationary activity
Facilities for daily living, such as a market in one of the segments, may record a higher Int. V, which means convenient access. Such segments might accommodate a higher level of necessary and social activities.

Hypothesis 3: Segments with a large amount of stationary activity
The ratio of commercial building use in a segment might, on the one hand, increase the activity because they also induce vendors. On the other hand, the activity might decline around public transportation hubs and public facilities, where the place function is controlled by the city administration to prioritize the walk-through movement.

Pedestrian-related information on the target cities is not integrated into the local government's data; therefore, as part of the research, we conducted a simple field survey to assess the actual conditions of stationary activity and Int.V of space syntax theory. We followed the definition of stationary activity by Gehl (2008) and classified the actual activities observed on-site into three categories: necessary activities (e.g., business or errand, waiting for a vehicle, meeting people), optional activities (e.g., socializing), and social activities (e.g., energy recovery, shopping, viewing other activities, meals). The acquired activity data were analysed as heat maps to test the hypotheses (see Figure 2.5). The additional building use data were input to derive the features of segments containing high-density activities. Conclusions were drawn, considering the differences and similarities between the two cities. Due to space limitations, we will only present the findings.

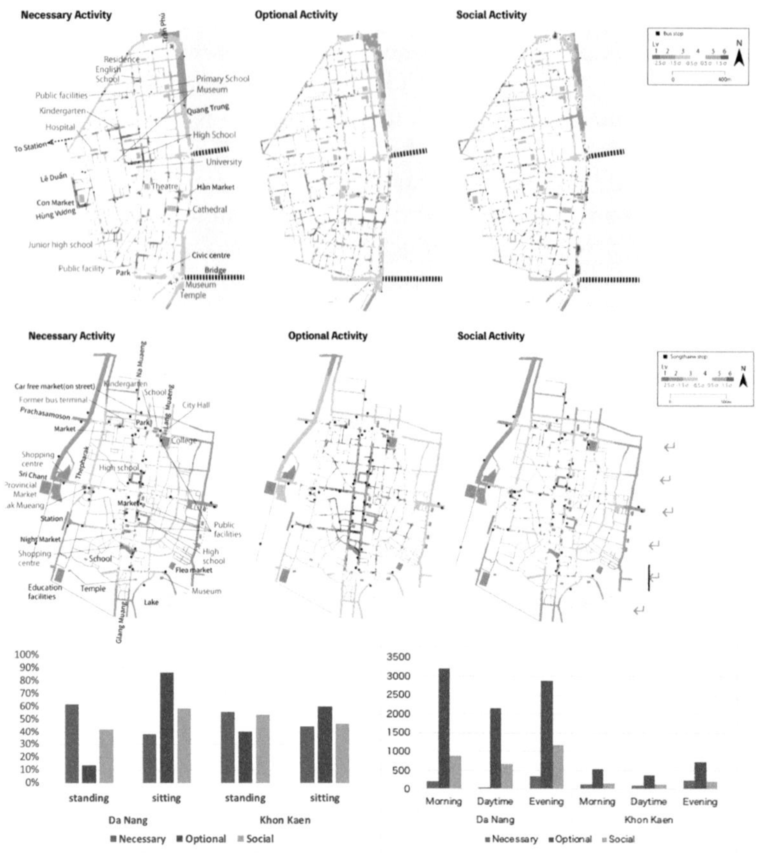

Figure 2.5 Heat maps of three activity classifications (Da Nang above, Khon Kaen City middle) and rate of posture (bottom left), number of activities in the three timeframes (bottom right) in the three activity classifications

Source: Miura et al. (2023)

2.2.4 Results

All three hypotheses on the spatial and timeframe distribution of pedestrian activities were partially rejected. Regarding Hypothesis 1, it is clarified that stationary activity, even for necessary activity, might be high during times other than daytime when the temperature is high. For Hypothesis 2, the study suggested that stationary activity and Int. V, which are related to walk-through activity, did not seem to have a generalizable relationship. For Hypothesis 3, the study found that high activity density and Int. V were sufficient conditions for the high ratio of commercial building use.

For Hypothesis 2, the results indicated the possibility of no correlation between walk-through demand and stationary activity demand. Meanwhile, the timeframe was found to be an influential variable related to stationary activity demand. The result of Hypothesis 1 and heat maps – highly necessary activities showing not only commuting destinations but also in front of religious monuments and markets – suggest the local pedestrian culture: the possibility of people planning errands in a more comfortable timeframe to visit the street-side facilities in Southeast Asian cities.

The results of Hypothesis 3 do not mean that the density of stationary activity was always high in front of commercial functions. However, commercial uses tended to be located along segments with a high degree of management difficulty, where traffic route potential and place function demand overlapped. In these places, government measures are needed to minimize pedestrian obstructions. However, the interviews in Khon Kaen indicated that current efforts and opinions on environmental improvement by individual stores have not been fully supported by the administration. The quality of sidewalk maintenance by the government is not adequate, even if the cleaning cost is met by fees collected from the occupants (i.e., shop owners and vendors). Policies that encourage businesses' participation in daytime street management, such as organizing local merchant associations to be influential partners with the administration, would be meaningful in the two cities and other urban areas in emerging Asian cities.

There are commonalities between the two cities. Both have the same balance of activities. At any time in the two cities, the volume is greater in the order of optional activities > social activities > necessary activities. Optional activities increase in the evening (see graphs in Figure 2.5). Sitting rates were higher for optional activities than for necessary activities (chi-square test < 0.01).

There are also differences between the two cities. The elderly are more often seated than other age groups (63%, chi-square test < 0.01). Total sitting posture is high in Da Nang, but more of the elderly seated are seen than other age groups in Khon Kaen. At evening, necessary activities increase in Khon Kaen and social activities increase in Da Nang. Throughout the daytime, the necessary activity rate is lower in Da Nang than in Khon Kaen. Concerning the concentration of high-density segments, Da Nang is relatively dispersed due to the existence of open spaces, including the river, which induces optional activities and social activities. In Khon Kaen, these activities are concentrated along the main street.

2.2.5 Conclusion

The findings provide insights into local street management when developing cities in Southeast Asia. There is a need to create a spacious open spaces strategy throughout the CBD rather than relying solely on policing individual businesses to improve the relationship between pedestrians and vendors. For spatial zoning, place demand occurs not only in the centre's main street but also in the inner city's alleys and outer areas, which attract social activity. There also should be a timeframe operation reflecting a shift in activities from daytime to evening in the urban pedestrian culture. For example, on the one hand, in a city such as Da Nang, where social activities, subject to similar spatial conditions to optional activities, are increasing, the layout of the place could be identical throughout the day. On the other hand, a layout change could be indicated in cities like Khon Kaen, where there are more rates of necessary activities and a segments of social activities tend to overlap with high-in-count segments of optional activities. The other point is the relationship between particular facilities and stationary activity. There is a high number of necessary activities in front of markets and schools. Transit stops are also surrounded by higher density meshes. Vendors are regulated in front of these public facilities, but there is pedestrian demand for necessary and optional activities. The point to remember for timeframe operation here is that while off-peak hours may be acceptable for the current walking environment, there may be still issues during peak hours. Especially at schools, which give rise to a concentration of necessary activity for commuting, measures such as setting up spaces within the block during peak hours can enhance the off-street place function while keeping the walk-through function.

Sidewalk improvement should be launched in front of public facilities, transit stops, and commercial use segments where higher density was observed. A segment with a greater proportion of commercial use tends to need more complicated management skills. Therefore, regulations and incentives for merchants should be coordinated. These include banning parking on the sidewalk and providing positive cooperation in managing street furniture and other loading spaces by compensating for the repair cost. To maintain inclusive place function, it would be helpful to create a place that considers users' posture (e.g., in Khon Kaen, where older adults tend to sit) and the overlapping demands of social and optional activities. The seating layout should offer appropriate zoning reflecting the trend of sitting users (e.g., having less space in front of product sales). When beginning to operate the operation guidelines, store owners who have relationships with stall owners and have high management motivation may be appropriate partners with the local government to manage the streets.

2.3 REDESIGN OF PUBLIC SPACE AND PEDESTRIANIZED STREET, THE EXPERIENCE IN CURITIBA, BRAZIL

Fumihiko Nakamura

2.3.1 Role of IPPUC and URBS

The city of Curitiba, the capital of the state of Paraná in the Brazilian Federation, is the fifth largest city in the Brazilian Federation, with a population of approximately 2 million people as of 2024. This city is very well known in the field of urban policy, with a wide range of practical experience. The Master Plan, developed in 1965 after a competition, is an excellent example of a basic concept that emphasizes people, environment, and history, and its basic form remains unchanged to this day.

The unique organizational structure responsible for the planning and implementation process in the area of urban policy in Curitiba is worthy of special mention. On the one hand, the Urban Planning Institute (IPPUC; Institute de Pesquisa Planejamento Urbano de Curitiba) was established directly under the mayor, and all urban policy proposals, as well as analyses and diagnoses of the current situation, are made at this institute. The IPPUC has a large number of professional staff members who have studied architecture and other fields, and they conduct a variety of surveys, analyses, and design proposals. The institute has long made full use of geographic information systems and has advanced management of land use and building data. On the other hand, the Urban Infrastructure Management Public company (URBS; Urbanização de Curitiba Sociedade Anônima) is responsible for the management of road facilities, parking management, and road traffic operations. The URBS is responsible for the management of the city's bus system, which is operated by 10 private operators in the city. The bus system and facilities are designed by the Urban Planning Institute, which also handles comprehensive transportation policy, including buses. The details of the routes and the operation of the facilities are the responsibility of the Urban Public Corporation, while bus operations, vehicle management, and crew management are divided among the various operators. The role of the Urban Corporation has been changed in detail several times, but the management of the bus service has remained unchanged from the beginning.

2.3.2 Structural Axes

Curitiba's urban policy is generally known for its structural axis concept, which has been realized step by step since 1974. As mentioned earlier, the plan was based on a bus system that incorporated an excellent management system with a clear division of roles between the public and private sectors by separating planning, management, and operation, and only along the main bus lines running on dedicated roads (which became known as BRT after about 20 years), low-rise buildings were built with the following features: Land use regulations have been introduced to guide the construction of high-rise residential complexes along the main bus lines, with the lower floors used exclusively for commercial functions such as restaurants, stores, and offices. This was later called TOD. Curitiba's TOD will be presented in Chapter 8, Section 4.

The structural axis is framed by three parallel multi-lane roads. The multi-lane roads at both ends of the axis are paired with one-way roads to handle arterial automobile traffic. The centre two lanes of the arterial road, the central spine, are dedicated to buses. The two lanes at either end are for intra-district traffic, and traffic volume is relatively low, with a speed limit of 30 km/h. The bus stops at the centre of the bus lanes. The bus lane in the centre of the street is assumed to be a level crossing between the bus stops and the roadside. Because of the low traffic volume and slow speed, it is considered safe. In recent years, humps have been introduced in pedestrian crossings to ensure safety.

The rows of stores and restaurants along both sides of the road can be described as similar to so-called transit malls (pedestrian-oriented commercial spaces where only public transportation is allowed to travel). In this sense, the bus lanes along the structural axis can be regarded as part of an excellent public space, although not necessarily beautiful in some respects.

2.3.3 Pedestrianization of Streets in Downtown

In the implementation of this master plan, the City of Curitiba has decided to convert the main street of the central district (Centro) into a pedestrian-only street at an early stage. This street is a continuous space of about 2 km in length and 15 to 30 m in width, traversing the central area of Curitiba from east to west.

According to interviews with local residents, the mayor initially failed to obtain the consent of shopkeepers along the street and forced them to install barricades to keep cars out of the street. A week later, the store owners attempted to remove the barricades with heavy machinery, but this information was leaked to the mayor's office in advance, and the mayor hurriedly held a sketch contest for elementary school students on a street with no cars. The store owners, who were about to storm the area with heavy machinery, gave up on removing the barricades in front of the elementary school students who were enjoying drawing. It can be interpreted as if there was a battle that could be interpreted as a human shield through a demonstration that was neither a consensus building through discussion nor a social experiment conducted peacefully.

However, there are many things to consider, such as the fact that they actually allowed the children to embody a car-free space, that they allowed the children, who are most expected to be pleased, to experience that space, and that they showed the community what it was like.

Later, as the street was being developed as a formal pedestrian-only space, many flower beds were installed (see Figure 2.6), earning it the nickname "Ohana Dori" (Flower Street).

2.3.4 Parks and Public Spaces in Downtown

In addition to the pedestrianization of the main street, many parks have been introduced in the central district to connect the flow lines and ensure a continuous public space. In addition, some of the parks also serve as bus stops, so to speak, in the form of open-air bus terminals.

The central district has prohibited the construction of new off-street parking lots since the 1970s. The intention was to ensure that bus access would be sufficient and to avoid worsening congestion on the roads leading to the central area. However, in 2004, the new mayor of Curitiba made a major policy shift, in which he lifted the parking lot construction freeze. As a result, many parking lots began to appear in the central district as buildings were reconstructed.

Figure 2.6 Pedestrian Street in Curitiba

Source: (Photos taken by Fumihiko Nakamura)

Furthermore, the Brazilian federal government positioned the automobile and oil-producing industries as key industries to be fostered, and as a national policy, introduced low-interest loans and reduced fuel taxes to encourage private car ownership and use. As a result, congestion on the roads in the central area worsened. The new mayor, elected in 2017, has reorientated policy and is now beginning to focus again on reducing car dependence, strengthening bus service, and so on.

The significance of the city of Curitiba's subsequent development of a public space system centred on pedestrian-only spaces in the central district during the early stages of motorization, and economic growth has been significant.

2.4 SHARED BICYCLE IN CHINA

Shengchuan Zhao

China was once called the "Kingdom of Bicycles" during the 1970s–1980s, and bicycles remained a major mode of urban transportation in most Chinese cities during this period (see Figure 2.7).

Since the 1990s, China has experienced a rapid motorization, and the road was getting more and more congested. As one of the countermeasures

Figure 2.7 Old downtown street in the 1990s, Wuhan, China

Source: (Photo taken by Shengchuan Zhao)

to alleviate road congestion, many local governments started to restrict or even prohibit the use of bicycles in urban areas, and changed bicycle lane into automobile lane. However, road traffic remained congested and air pollutions were getting worse.

The role of bicycle has been rediscovered since the 2010s. Shared bicycle has been quickly and widely developed in China due to its convenience, flexibility, and affordability. This section reviews the bicycle policy in the past decades, then describes the development of bicycle sharing systems. This is followed by an introduction of a number of bicycle projects. The section concludes with a summary of the major bicycle issues in China.

2.4.1 The Evolution of Bicycle Policy in China

Since the implementation of "Reform and opening up" policy in 1978, China has enjoyed a rapid economic growth, the number of private cars has increased very quickly, and the lifestyle has changed dramatically. Under this background, the bicycle industry and bicycle transportation have gone through the growth period, booming period, recession period, and rediscover and innovation period described in Table 2.1.

Table 2.1 The evolution of bicycle policy in China (1949–present)

Key indicators and policies	*Growth period (1949–1978)*	*Booming period (1979–1991)*	*Recession period (1992–2009)*	*Rediscover and innovation period (2010–present)*
Bicycle modal share	High	High->Low	Low	Low->Slightly high
Bicycle policy	Strong planned economy. The bicycle price was regulated based on the per capita disposable income, and there were strict quota restrictions on the supply of bicycles and an annual license tax.	Affordable bicycle price regulated by the government, employee's bicycle commuting subsidy, bicycle lane development.	Restrictions or even prohibitions on the use of bicycle, charging bicycle annual tax, rebuilt bicycle lane into automobile lane or merges the bicycle lane with pedestrian lane.	Rediscovering the role of bicycle, introduction of public bicycle systems in major cities, rapid development of bicycle sharing systems funded by private sectors.
Urbanization rate (%) (year)	10.6 (1949)->17.9 (1978)	19.0 (1979)->26.9 (1991)	27.5 (1992)->48.3 (2009)	50.0 (2010)->65.2 (2022)
Car ownership (Car/1000 people) (year)	1.41 (1978)	1.61 (1979)->5.23 (1991)	5.90 (1992)->47.06 (2009)	58.18 (2010)->220.89 (2022)
Urban travel distance	Short	Short	Short->Long	Long
Note	Bicycles were luxury goods. Owning the "Big Three" (a watch, a sewing machine, and a bicycle) was the dream of many families in the 1970s.	Bicycles were household necessities. Bicycle ownership reached approximately 2 bicycles per household. Called as "The Kingdom of Bicycles".	Rapid motorization. Car became a symbol for an individual or a family's social status.	Beijing Olympic Games, Shanghai World Expo and Guangzhou Asian Games, Emergence of "Internet+" and sharing economy models.

Note: Key indicators are cited from National Bureau of Statistics of China (2023).

Table 2.2 Bicycle sharing systems in China

Provider	*Business model*	*Examples*
Public transportation agency	Government-led, company operates	Hangzhou, Guangzhou, Shenzhen
Private company	Company runs on advertising revenue	Beijing, run by Fortune Public Bicycle; Hello Bike; DiDi Bike; and Meituan bike
Public-Private Partnership company	Company-led, government aid; Government buys the services from company	Shanghai, run by Shanghai Forever Group (Bicycle maker)

2.4.2 Bicycle Sharing Systems

Bicycle sharing systems can be roughly grouped into three types, as shown in Table 2.2. The upcoming sections will introduce the first two types of bicycle sharing systems.

2.4.2.1 Hangzhou Public Bicycle System

The first large-scale IT-based bicycle sharing system in China was launched in Hangzhou, a historical and tourist city in China, on 1 May 2008. Under the "government led, company operates" business model, the system is planned, invested, and built by the government, and is operated by Hangzhou Public Bicycle Transportation Service Development Co., Ltd., a public agency belonging to Hangzhou Public Transportation Group Co., Ltd. Hangzhou city government invests in the infrastructure construction, and the company is responsible for the operation. The system is integrated into public transportation system in Hangzhou, and has been expanded throughout the city.

Table 2.3 outlines the key statistics for the system in the initial period. By the end of June 2024, the system has owned 5,635 service points and 146,000 public bicycles, with the daily maximum number of rentals reaching more than 473,000, the cumulative number of rentals exceeding 1,356 million, and the rate of free use (for the first hour of rental) reaching 98%. The service points are located at public transportation terminals, large residential communities, public service buildings, sightseeing areas, and schools and colleges. The bicycle service radius is approximately 100–135 meters.

The system is constantly making various efforts to provide more convenient services, such as charging, beverage vending, baby stroller rental, and express delivery storage services. It allows riders to return bicycles overnight or staggered return without charging additional fees. The system covers any injury that occurs during proper bicycle use and provides third-party liability insurance. The system has received many domestic and international awards and was selected as one of the five best cycling practical cities in the world by the C40 Cities Climate Leadership Group at the Hangzhou International Day in 2021.

Table 2.3 Key statistics for Hangzhou Public Bicycle System

Date	*No. of shared bicycles*	*No. of service points*	*Service areas*	*No. of average daily rentals*	*No. of average rentals per bicycle*
May 2008	2,500	61	West lake scenic area, Central urban areas (Western and Northern sides)	5,000	0.93
May 2009	17,000	800	Central urban areas (six districts including scenic area, West lake, Gongshu, Jianggan, Shangcheng, and Xiacheng districts)	99,798	5.6
May 2010	50,000	2,000	Central urban areas + Xiaoshan district	301,647	6.04

Source: Shi et al. (2011).

Figure 2.8 Hangzhou public bicycle, China

Source: (Photo taken by Chao Liu)

2.4.2.2 *Bicycle Sharing Systems in China*

With the rise of the new economic models of "Internet Plus" and sharing economy, bicycle sharing systems emerged in China in the mid-2010s. Ofo, a shared bicycle pioneer, started in Beijing in 2014 by providing travel tools on campus. Mobike was launched in Shanghai at the same time. Unlike Hangzhou public bicycle system, shared bicycles are owned and operated by private companies, which involved no urban planning, but received huge financings from venture capitalists. Within a short time, dozens of companies joined the market. Each has its own app and a fleet of bicycles in a distinctive colour.

Different from most programmes abroad, shared bicycles in China did not require docking stations and riders could drop them off anywhere. The bicycles are embedded with a GPS (Global Positioning System) chip, allowing riders to find them using a smartphone app. After registering and making a refundable deposit, riders could unlock a bicycle simply by scanning the QR code on the frame. The payment is made electronically.

Bicycle sharing in China has experienced exploration, outbreak, and recession periods, and now enters a rational development stage, as described in Table 2.4.

Since 2018, bicycle sharing market has learnt bitter lessons from the excessive competition without profitable business models, and started a new round of development. Many small companies went bankrupt, while a few companies have been acquired by technology giants. For example, Mobike, a leading company until 2018, was acquired by Meituan in April 2018. Ofo was caught in the "refund deposit" predicament. Hello Bike chose to cooperate with Ant Group, by leveraging Ant's substantial financial strength and market reputation; it became the largest bicycle sharing company in China. DiDi Bike, a bicycle sharing service from Didi Chuxing (DiDi), was

Table 2.4 The development of bicycle sharing systems in China

Period	*Key characteristics*
Exploration period (2014–2015)	Ofo, a pioneer startup company, emerged on campus. Mobike started its service in Shanghai.
Outbreak period (2016–2017)	The number of shared bicycles increased from 2 million to 23 million. The number of registered riders increased from 19 million to 221 million. 70+ companies were established. Overseas business started.
Recession period (2017–2018)	Excessive competition and weak government regulation caused many companies to go bankrupt. The number of companies reduced to less than 10 in 2018. The two giant companies, Mobike and Ofo, were also shrinking.

established in January 2018. It utilized DiDi's capital and management model to quickly become a big giant in the market. Currently, Hello Bike, DiDi Bike, and Meituan Bike are the top three companies with the most registered riders in China.

Despite some ups and downs in the past decade, China's bicycle sharing market is growing steadily, and the number of shared bicycle users increased from 310 million in 2017 to 460 million in 2022. It is estimated to exceed 500 million people in 2024.

2.4.3 Typical Bicycle Projects

A number of bicycle projects have been conducted to enhance the bicycle use in China. Xiamen bicycle skyway is one of them which receive the world attention. The project is the first elevated bicycle road in China and the longest bicycle bridge in the world. It is 7.6 km long and runs along underneath the city's existing overhead BRT skyway in the central part of the city. It covers the city's major residential areas, commercial centres, public service buildings, and schools. There are 11 entry points which connects with 6 BRT stations and 3 pedestrian bridges. The bicycle skyway is designed along both sides of BRT, two lanes on each side; the net width is 2.5 m. The design speed is 25 km/h.

Xiamen bicycle skyway won the "Livable Cities" category in the Danish Design Award 2019, as it is recognized as "having a positive effect on the city environment and ecology".

Figure 2.9 Xiamen bicycle skyway, China

Source: (Photo taken by Ping Yan)

2.4.4 Bicycle Issues in China

Bicycle transportation and bicycle sharing systems in China have undergone remarkable transformation in the past decades. China has made a lot of efforts towards a sustainable bicycle development; however, China is still facing many challenging issues as follows.

(1) Insufficient policy support: Robust policies at both the central and local government levels are essential to shift away from the prevailing "car-centric" mindset and to promote bicycle-friendly urban planning. The public-private partnership model for bicycle sharing systems should also be encouraged to enhance service efficiency and sustainability.
(2) Inadequate bicycle infrastructure: Bicycle lanes are often insufficient and poorly connected, as the majority of road space continues to be allocated to private cars. Moreover, bicycle parking facilities are severely lacking and fall short of meeting actual demand.
(3) Weak law enforcement: Bicycle lanes are frequently occupied by illegally parked private cars. Disorderly bicycle parking is widespread. Safety regulations such as mandatory helmet use and the requirement for bicycle lights at night are often neglected and should be strictly enforced. Additionally, the issue of speeding electric bicycles must not be overlooked.
(4) Limited public awareness: The contributions of bicycles to achieving Sustainable Development Goals (SDGs) are not widely recognized. Greater efforts are needed to raise public awareness and actively promote cycling as a sustainable mode of transportation.

2.5 CONCLUSION

Mihoko Matsuyuki

In this chapter, the roles, usage, and spaces for walking and cycling as active transportation are discussed. Section 2.1 examines the roles and values of active transportation and its spaces. In developed countries, the value and role of active transportation spaces have been revalued, leading to their expansion. Pedestrian spaces are not only used for traffic of pedestrian but also serve various purposes, such as supporting local businesses and fostering community activities. In emerging countries, while the quality of pedestrian spaces is insufficient, these spaces are actively used for dining, communication, and recreation, making the neighbourhood lively and vibrant. However, some emerging cities are increasingly banning non-passage uses, such as street vending, and reallocating pedestrian spaces for cars, moving in the opposite direction compared to developed countries.

Section 2.2 focuses on sidewalks not as "spaces" for through traffic but as "places" to linger, reporting the results of surveys conducted in Khon Kaen (Thailand) and Da Nang (Vietnam). The survey revealed that optional activities and social activities increase in the evening, leading to crowded pedestrian spaces during peak hours. Necessary activities are concentrated around markets and public facilities, highlighting the need for balance between passage and stationary activities in these areas. While street vendors enhance community ties and economic activities, a lack of proper management poses safety risks and obstructs pedestrian traffic.

Section 2.3 introduces Curitiba's public space redesign. Along Curitiba's structural axes, developments combining commercial facilities and residential buildings have progressed, creating a transit-centred environment. Additionally, the main street in the central area has been converted into a pedestrian-only road, making it an important public space for residents and tourists.

Section 2.4 outlines the development of bicycle sharing in China and its associated challenges. Bicycle-sharing services started in China around 2014, experienced explosive growth around 2016, and later saw a decline due to excessive competition and weak regulation. Subsequently, major technology companies took over the services, leading to a period of stable growth. This section proposes some countermeasures against the issues of bicycle sharing system. To address the issues, government policy support, infrastructure development for bicycle lanes, stricter regulations against illegal use of bicycle lanes by cars and against improper bicycle usage, and raising public awareness of bicycles' contributions to the SDGs are essential.

It is evident from the cases of Curitiba and China that the value of active transportation is recognized in emerging countries too. Securing space for active transportation has promoted its use and curbed car dependency. In China's case, modernization through ICT integration with bicycle-sharing has significantly revived bicycle usage, which had previously been replaced by automobiles. However, ongoing motorization is still hindering the utilization of active transportation and pedestrian spaces.

Thus, the value of active transportation and its associated spaces is gradually being revalued in emerging countries. However, challenges remain, such as managing the balance between "spaces" and "places" functions in pedestrian areas during certain times in Southeast Asia; and addressing issues like improper bicycle parking, riding without lights at night, and speeding in China. Proper management and regulation are crucial to encourage active transportation usage and balancing traffic and various activities in pedestrian spaces.

Additionally, further development of pedestrian spaces and infrastructure for bicycle is necessary, as they are key to enhancing the environmental and social sustainability of cities. However, merely developing these spaces is insufficient; maintenance after implementation and proper usage regulations are essential for their success.

REFERENCES

Gehl, J. (2008). *Life between buildings: Using public space* (6th ed.). Island Press.

Lah, O. (2019). Chapter 7 – sustainable urban mobility in action. In O. Lah (Ed.), *Sustainable urban mobility pathways* (pp. 133–282). Elsevier. doi:10.1016/B978-0-12-814897-6.00007-7

McMillen, B. (2003). Making pedestrian facilities more usable and safer for all. In R. Tolley (Ed.), *Sustainable transport* (pp. 298–313). Woodhead Publishing. doi:10.1016/B978-1-85573-614-6.50030-6

Miura, S., Yoshida, M., Nakamura, F., Wongwiriya, P., Tanaka, S., & Ariyoshi, R. (2023). Spatial and timeframe distribution of diverse pedestrian activities in the street network of southeast Asian developing cities. *Journal of Asian Architecture and Building Engineering*, *23*(1), 386–404.

Monheim, R. (2003). The role of pedestrian precincts in adapting city centres to new lifestyles. In R. Tolley (Ed.), *Sustainable transport* (pp. 326–338). Woodhead Publishing. doi:10.1016/B978-1-85573-614-6.50030-6

National Association of City Transportation Officials. (2016). *Global street design guide*. Island Press.

National Bureau of Statistics of China. (2023). *China statistical yearbook 2023*. China Statistics Press. (in Chinese)

Shi, X., Cui, D., & Wei, W. (2011). Investigation of the relationship between using and planning of public bicycle in Hangzhou. *Urban Studies*, *18*(10), 105–114. (in Chinese)

Chapter 3

Automobile, Motorcycle, and Freight

3.1 TRANSITION FROM BICYCLE AND MOTORCYCLE CITY TO CAR CITY

Fumihiko Nakamura

3.1.1 From Walking to Vehicles

In terms of the main means of transportation on roads, not only in emerging countries but also in developed countries, history has seen a transition from the era of walking, to the horse-drawn carriage, to the bicycle era, to the motorcycle era, and then to the automobile era. In the automobile era, there were horse-drawn railroads, in which horse-drawn carriages were pulled by carriages on tracks laid on the road, and streetcars, in which overhead wires were also installed on the road and ran on the road. In many developed cities, streetcars were first introduced and then removed as the number of automobiles increased. Some cities have recently reintroduced them.

3.1.2 Beginning of Motorcycles

Now, the transition from walking to bicycles and motorcycles, and then to automobiles, seems to be common in many emerging cities, but this is not always the case. We will discuss more about this in the following sections. It should be noted that the points raised here are also important perspectives for future urban transportation policies.

There are various definitions of a city, but for simplicity's sake, let us define a city as a place where many people live in relatively close proximity and at a relatively high density, and where a variety of economic activities are conducted. It is not difficult to imagine that if the area (city) where residences and places of economic activity (stores and offices) are clustered together is within, say, 15 minutes or about 1 km, it is sufficient for people to move around and for goods to be transported on foot or by cart.

DOI: 10.1201/9781003467403-4

Eventually, as the economic power of a city increases, more and more people will seek to participate in economic activities and reside in that city. Therefore, in many cases, cities expand spatially. In exceptional cases, when it is difficult to spread out spatially, if there are no technical constraints, living spaces spread out in the sky direction; that is, multi-story or high-rise residences emerge.

As cities spread out on a flat surface, the distances people travel and the distances they carry increase. This is where the need for transportation tools other than walking and carts – that is, other than human power – arises. Before the invention of bicycles and motorcycles, animal-powered means of transportation were the main means of transportation; after the 19th century, when bicycles and then motorcycles became popular, they were also used. Thus, cities incorporating bicycle and motorcycle travel and transportation are born. However, geographical, economic, and trade conditions are involved. This means that urban expansion alone cannot be explained.

First, let me explain the geographical situation. Bicycles do not spread easily in areas with significant topographical differences in elevation. Cities themselves do not expand evenly. In most cases, cities expand within the range of flat terrain, but in cities that have a different history, bicycles do not spread so much, and the transition to motorcycles, which can travel on slopes without problems, or automobiles seems to follow.

Next, let us discuss the economic situation. Bicycles, and even more so motorcycles, are expensive to purchase and use. Thus, even in cities, people with low incomes are not able to take advantage of them. In cities where income inequality is very high, some people spend many hours walking long distances, which means that they cannot afford transportation. This can also be explained by the fact that people have low incomes and low time value, so they accept travel methods that take longer to get around. Thus, even as cities expand, not all journeys over a certain distance can be substituted by transportation devices such as bicycles or motorcycles. Naturally, motorcycles are more expensive than bicycles and require more gasoline, so some cities go through stages of walking to bicycles and then bicycles to motorcycles. In cases where it takes time for motorcycles to spread due to the circumstances described next, a bicycle-centered period may exist for a long time. Outside of China, Mandalay, the ancient capital of Myanmar, is known as a city that experienced a bicycle-centered era.

Yet another circumstance is what might be called the trade situation. Motorcycles cannot be produced anywhere in the world. On the one hand, the production of engines, in particular, requires a certain level of technology, so they can only be produced in a limited number of countries. In countries with production capacity, the spread of motorcycles seems to have been rapid. On the other hand, countries without production capacity had no choice but to import them. This is where tariffs and other trade-related issues come into play. As for gasoline, which is used as fuel, if a country can produce crude

Figure 3.1 Bicycles in Mandalay in 2010*

Source: (Photo taken by Fumihiko Nakamura)

oil and refinery oil on its own, it is still a burden on trade. In some cases, it is an extremely expensive vehicle. However, this situation can change rapidly with the revision of tariffs and the establishment of new production bases and manufacturing plants. This is the reason why motorcycle registrations vary from country to country over time, and why the growth is not linear.

3.1.3 From Bicycles and Motorcycles to Automobiles

Now we turn to the case of a bicycle- or motorcycle-centered city transitioning to automobiles. Basically, as cities grow larger and incomes rise, they replace bicycles with motorcycles and then motorcycles with automobiles. As a city's economic activity increases and secondary industries are promoted, the amount and distance of goods transported will increase, and the use of cars, trucks, and other vehicles for logistics will become significantly more widespread. In cities with canals and rivers, transportation by boat coexists, but as factories are located inland, trucks and other vehicles become the main means of transportation.

However, observing the changes in Hanoi, Vietnam, for example, is not as simple as the aforementioned. In the city center, where motorcycles have been the main means of transportation for many years, building corridors and side streets within city blocks (meaning sites or blocks surrounded by roads) were built on the premise of motorcycle driving and storage, and they cannot physically accommodate automobiles, which are much larger than motorcycles as they are. The streets themselves can be redeveloped, but in that case, considerably taller buildings will have to be constructed to maintain the previous high density. As for roads, a good stock of roads was prepared during the colonial period, and the number of excellent roads has been increasing with the support of developed countries in the subsequent expansion of the city. It is extremely difficult to replace all motorcycles with cars, whether for passing, stopping, or parking.

Given the constraints of urban space, it is clear that considerable challenges exist in converting from motorcycles to automobiles from the perspective of maintaining population density and the volume of urban activity.

This is one of the reasons why there are as many motorcycles as ever in Hanoi. Naturally, there is the issue of income. Furthermore, there is also the feeling among citizens that they will own and use a car as a status symbol when their income rises, and that riding a motorcycle over a certain distance (said to be several kilometers) is physically demanding, especially for women. Also, in countries where land is publicly owned, it is not easy to acquire land for parking. However, this does not mean that motorists are allowed to park their cars on the sidewalks or roadside areas in the same way that motorcycles are parked on the sidewalks. The problem of on-street parking in the urban center of Vientiane, Laos, is also connected to this point.

In this way, the change from motorcycles to automobiles is not a simple substitution, although it is based on urban expansion or growth. It is necessary to understand the reality of road space, urban space, building construction, regulatory control, income level, and the desire to display status.

3.2 MOTORCYCLE SAFETY ISSUE (ENGINEERING, EDUCATION, AND ENFORCEMENT)

Fumihiko Nakamura

3.2.1 General Points for Traffic Accident Prevention

Motorcycles became popular in many Asian cities. As mentioned earlier, there are differences among countries in the timing of diffusion, especially due to trade conditions. For example, while the spread of motorcycles was quicker in Vietnam, Taiwanese cities influenced by mainland China, Thailand (which had a production base from early on), and Indonesia were a little behind, followed by the Philippines and other countries.

3.2.2 Engineering, Education, and Enforcement

Motorcycles are two-wheeled and less stable than four- or three-wheeled vehicles. As a result, they are said to be more prone to traffic accidents. The problem is so serious that the founder of Honda Motor Company, one of the driving forces behind the popularization of motorcycles in Asian countries, donated his personal fortune to establish the International Association of Traffic and Safety Sciences (IATSS) to reduce motorcycle accidents in those countries (IATSS, 2015).

Thanks in part to the Society's initiatives, there have been multifaceted efforts to combat motorcycle traffic accidents. Today, it is appropriate to understand them from the perspective of the 3 Es: Engineering, Education, and Enforcement.

The first is the Engineering aspect, which is mainly about motorcycle vehicle engineering and road traffic engineering. Motorcycle vehicles cannot be changed significantly due to their simple structure. Rather, measures on the road management and traffic management side fall into this category. Examples of such measures include setting up lanes and stop lines for motorcycles and devising signal control. In most cities, including those in developed countries, road design and traffic management techniques are based on the assumption of four-wheeled vehicles. However, when four-wheeled vehicles and motorcycles are mixed in a road space and lanes are designed with only four-wheeled vehicles in mind, various intersections of traffic lanes inevitably occur, increasing the risk of accidents. This increases the risk of accidents. Physically separating the lanes for motorcycles (and bicycles), as in some Chinese cities, will also increase the effect of accident reduction.

Next, let's look at education. It is hard to imagine in developed countries, but there are many countries that do not require a motorcycle driving license or that do require one, but the procedure for obtaining one is very simple. In many cases, people feel that they will be able to drive a motorcycle when they are a little older, in the same way that a small child can learn to ride a bicycle with practice when he or she is older. There is no training course for obtaining or renewing a driver's license, and there is no education on the laws and regulations concerning road traffic. There are no educational materials or educational systems that can help students accurately understand the laws and regulations at an educational level where they cannot read or comprehend texts to begin with. Even though many junior high school students and above drive motorcycles, there is no traffic safety education in schools either. Fortunately, many of them have good motor skills, a good sense of balance, and excellent reflexes, and they are able to avoid the risk of traffic accidents in some cases. In Japan and many other countries, there is a good track record in driver licensing and road safety education. There is also a lot of knowledge about road safety education in school education. Therefore, it goes without saying that it is necessary to have an appropriate driver's license system and, at the same time, to provide thorough traffic safety education. It is said that the meaning and value of life differ from country to country due to differences in religious backgrounds and economic conditions, but there is nothing better than not to die. Efforts to eliminate traffic accidents and traffic deaths are an essential task.

Finally, there is enforcement. Even if laws and regulations are in place, it goes without saying that compliance with those laws and regulations and enforcement of violators are important. Proper enforcement requires training of traffic police officers. Careful design and operation of the system is a must, as it can be a breeding ground for bribes.

3.3 MOTORIZATION AND ROAD DEVELOPMENT

Toshiyuki Okamura

3.3.1 Car Ownership in Emerging Market

Motorization is a phenomenon that combines the expansion of car ownership and increased use of cars. It leads to changes in lifestyle and changes in the location patterns of cities, which in turn accelerates motorization and pushes society further toward dependence on cars. In general, on the one hand, in emerging countries, the per capita car ownership rate is higher in large cities than in small cities and rural areas. This is not because people in large cities have a lifestyle that requires more cars, but because the income level in large cities is relatively high and there are more people who can afford to buy cars. On the other hand, in developed countries, smaller cities and rural areas, where the level of public transportation services is relatively low, are more dependent on cars. To put it very roughly, once GDP per capita exceeds US$ 10,000, the difference in car ownership rates between large cities and smaller cities and rural areas becomes smaller. Of course, this is largely dependent on the income gap between cities and rural areas, and the level of public transportation services in cities.

In addition to income levels, the progress of motorization is also related to the costs of owning a car, particularly the car-related tax system. The taxes on cars can be broadly divided into those on the purchase and ownership of the vehicle itself, and those on fuel. In many emerging economies, taxes on the purchase and ownership of cars are often set at high levels. In particular, tariffs on imported cars are often set at high levels. The main reasons for this are that cars are considered a luxury item, and that a certain amount of tax revenue can be expected. In countries where cars are manufactured or assembled domestically, taxes may be set at a lower rate to promote and protect domestic industry and stimulate the economy. In addition, in order to promote environmental measures and industry, preferential tax rates may be set for electric and hybrid cars. In fact, when taxes on car purchases are lowered, the number of cars sold increases and the car ownership rate rises. On the one hand, in oil-producing countries, fuel taxes are often set at a lower rate or even subsidized. On the other hand, non-oil producing countries generally impose certain tax rate. Even so, fuel tax rates in emerging countries tend to be lower than that in developed countries. Since the price of fuel itself is almost the same across the globe (e.g., the price of gasoline was approximately US$ 0.9 per liter in 2024), people in countries with relatively low incomes feel a very heavy burden when it comes to fuel. For this reason, it is not realistic to impose a tax rate higher than that in developed countries, especially European countries.

The price of cars is basically the same across the globe, and in that respect, it is not easy for people in countries with relatively low incomes to buy their own cars. However, when it comes to vehicles, there is the option of buying a used car as well as a new one. In emerging countries, few used cars are in good condition, and they are mainly imported from developed countries. Depending on the country, there are often certain restrictions and rules on the import of used cars, aiming to protect domestic industry and eliminate older vehicles with poor environmental performance. Depending on the condition, these used cars can be purchased for half or even a fraction of the price of a new car, and even when taxes are taken into account, they are still at a price level that is affordable for people in emerging countries. For example, in many African countries, many of the cars on the road, especially private cars, are used cars. In urban areas of African countries, the number of privately owned cars per capita is high in comparison to the relatively low GDP per capita, but the reason for this is that, in the face of inadequate public transportation and concerns about public safety, it is now possible to obtain cheap imported used cars. The United Arab Emirates is a major transit point for used car trading, mainly for Middle Eastern and African countries, and many used cars are traded here.

3.3.2 Road Development in Emerging Countries

There is a strong relationship between the progress of motorization and road development. The spread of automobiles promotes road development, and vice versa. In general, in emerging countries, the larger the city and the more developed the economy, the more of an issue the absolute lack of road infrastructure becomes. In particular, in the cities of emerging countries, the lack of road infrastructure leads to an increase in various social costs and can also be a factor that hinders economic development.

In many cities in emerging countries, both private and public transportation, as well as non-motorized modes such as walking, are highly dependent on roads. Even in cities in emerging countries where urban rail has been introduced to a certain extent, the modal share of rail is only around 10% at most when looking at the city as a whole. Roads are the most basic infrastructure for urban transportation, be it in emerging or developing countries.

Let's look at what role roads play. The functions of roads can be divided into two categories: traffic mobility, which is for the passage of vehicles and people, and land access, which is for entering and exiting along the roadside.

The former is typified by trunk roads, where the emphasis is on efficient passage of vehicles, and the latter is typified by streets in commercial and residential areas, where the emphasis is on functions for pedestrians, bicycles, public transportation, loading and unloading, and parking. The basic principle of traffic engineering is that different roads should share these functions, rather than having a single road take on all of them.

In the first instance, in the roads of the urban areas of emerging countries, particularly in the old city areas, one road is responsible for a mixture of these functions, and it is common to see a jumble of through traffic and access traffic, cars passing through and parking and unloading, and cars and pedestrians mixing together. As a result, the road space, which is already in short supply, is used even more inefficiently, and there are also major problems with road safety and the local environment. The road space for pedestrians is also extremely poor. The function of roads is not just for traffic and access, but also as urban space. Despite the fact that they are responsible for various urban activities, such as pedestrians, bicycles, and informal economic activities, in this situation, the vehicles of the minority of high-income car owners make up the most of the road space, and this is a problem from the perspective of social equity.

Furthermore, in emerging countries, the rapid expansion of urban areas is underway due to population growth and the increase in income levels as a result of economic development. In the process of urban area expansion, the development of road networks is essential. As cities become more suburban, population and various facilities naturally locate along the arterial roads radiating out from the city center, forming roadside urban areas. And the mixture of road functions and resulting traffic chaos that occurs in the old city center described earlier also occurs in the suburbs. It is important to coordinate road development and planning with land use planning and control, not just road planning, in the face of motorization. At the same time, in the development of arterial roads that mainly handle through traffic, it is important to develop ring roads to prevent traffic that does not originate or terminate in the city center from entering the city center, and planning and implementation are required in advance of the expansion of urban areas and the increase in vehicle traffic.

Some cities have developed ring roads and roads leading into the city center using access-controlled urban expressways (e.g., Shanghai, Bangkok, Manila, Jakarta, Nairobi). Many urban expressways are elevated highways that make use of the space above existing roads and rivers, and are built in densely populated city centers where it is difficult to add road infrastructure despite the need for it. Although they have been criticized for their impact on the urban landscape and the environment, they have dramatically reduced traffic congestion in city centers. They are built and operated using a toll road system. In addition to urban expressways, there are many cases where toll roads are built and operated in specific sections of cities (bridge and tunnel sections). In addition to public funds, the funding for road development in emerging cities also includes the use of ODA, loans from Multilateral Development Banks (MDBs) such as the World Bank and the Asian Development Bank (ADB), and in some cases, private sector funding through Public-Private Partnerships (PPPs). In particular, PPPs based on concession agreements with private companies have also been introduced for access-controlled toll roads. Determining how to set the level of tolls is not easy, as it involves deciding how much of the city's road traffic should be diverted to the urban expressway and how much the users are willing to pay, as well as ensuring the profitability of the expressway business (especially when private funds are involved).

From the perspective of road development funding, we will again touch on the automobile-related tax system. The tax revenue mentioned previously is also used for road development. In some countries and cities, the use of tax revenue is limited to road development. As the number of cars increases, so does tax revenue. Thus, from the perspective of satisfying the need for road maintenance and striking a balance between benefits and burdens, this tax system is excellent. In many developed countries, these taxes are the main source of funding for road development, and in some cases they are also used to encourage the reduction of car ownership and use, mainly from an environmental perspective, or to fund the development of public transportation within cities. For this reason, on the one hand, in developed countries where road development is well advanced and expenditure on road development is relatively small, tax revenue from automobiles tends to be the same as, or even exceed, the cost of road development. In emerging countries, on the other hand, these tax systems are often seen as an important source of revenue rather than as a means of discouraging the ownership and use of automobiles. As road development is also required to cope with the increase in the number of vehicles, the necessary road development costs are almost the same as, or even exceed, the revenue from automobile-related taxes in many countries, but some countries cannot spend sufficient amounts on road development as a result of the low rates of automobile-related taxes. In general, it is necessary to note that automobile-related tax rates and tax amounts (including subsidies) tend to be a political issue, and that tax rates and tax amounts are often set too low from the perspective of the necessity of road development and the social costs of automobiles.

3.3.3 Motorization and Road Development: Challenges for Emerging Countries

Based on the aforementioned analyses, from the perspective of motorization and road development, the challenges for emerging cities can be summarized as follows. First, a pricing system and level of charges, including a tax system based on an appropriate balance of benefits and burdens, should be established for the ownership and use of automobiles. In emerging economies, where steady road development is required, there should be a more explicit linkage between the level of automobile-related charges and the financial resources for road development. Second, the balance between improving the quantity and quality of roads and developing non-road urban transportation infrastructure is an important issue. It is impossible for roads to satisfy all traffic demands. If urban transportation infrastructure is excessively weighted toward road development, it will accelerate the growth of private vehicles and lead to the decline of public transportation, which will worsen the urban environment and contribute to mobility divides in society. Third, the priority should be given to road development for pedestrians, non-motorized transportation (NMT), and public transportation. Ensuring the safety and comfort of NMT, including walking, which is responsible for access and egress to public transportation, is the key to promoting the use of public transportation and thereby curbing car use, which could curb spending on endless road development.

3.4 CAUSES OF TRAFFIC JAM

Shinji Tanaka

Traffic congestion (or traffic jam) is a common issue in many cities of the world, as shown in Figure 3.2. It occurs when the traffic demand exceeds capacity. Here, traffic demand is the volume of vehicles that travel. Capacity

Figure 3.2 Traffic congestion

Source: (Photo taken by Shinji Tanaka)

is the ability of the road that facilitates these vehicles, and it can be regarded as supply in the context of economics. Therefore, it is a matter of demand and supply, basically. Traffic demand varies depending on time of a day, day of a week, and season of a year. Capacity is a relatively stable value compared with traffic demand, but still it can vary by road geometric factors, weather conditions, characteristics of road users, and so on. Based on the combination of these factors, congestion happens when the traffic demand becomes larger than the capacity. This is the basic mechanism of traffic congestion. Therefore, causes of traffic congestion can be categorized into the demand side (excess of demand) and the supply side (shortage of capacity). These two aspects are discussed, respectively, in detail in the following subsections.

3.4.1 Causes on Demand Side

In emerging countries, traffic demand has been increasing drastically due to the population increase, sharp economic growth, and rapid motorization. Economic growth increases both the number of vehicle ownership and the vehicle miles travelled (VMT), which is the largest contributor to increased traffic volume.

Not only the volume itself, but this drastic change produces a big discrepancy of road users, which results in the wide variety of vehicles on the road. There are some luxury, high-performance cars on the one side, and old, worn-out cars on the other. Even human-driven vehicles which are generally called NMT exist in the same road space. There are also a lot of overloaded buses and trucks. Such a variety of vehicles produces big differences of vehicle speed, acceleration, maneuver, and so on, and such complexity acts as a substantial increase in traffic demand compared with the case of a single type of vehicle (Ramanayya, T. V., 1988).

The increase in vehicle size that accompanies economic growth is another factor contributing to the increase in traffic demand. Former motorcycles are replaced by passenger cars, or smaller passenger cars are replaced by larger ones, then the area occupied by vehicles on the road increases, and traffic congestion is more likely to occur. Since a private car is a symbol of social status in emerging economies, the trend toward larger vehicles is more prominent in emerging economies than that in developed economies.

In recent years, pick-up and drop-off of children commuting to and from school by their parents has also become a source of traffic congestion. In emerging economies, on the one hand, the desire to provide their children with a good education is growing along with economic growth, and an increasing number of families want their children to enroll in good schools even if they are far away from their home. On the other hand, the environment is not conducive for children to commute to school by themselves, due to undeveloped sidewalks and a lack of school buses and public transportation. As a result, parents drive their children to and from school by their own cars. These vehicles gather in the vicinity of schools, often causing traffic congestion as they park and stop to let their children in and out of school (Lu, M. et al., 2017).

3.4.2 Causes on Supply Side

The road facilities that determine the capacity are generally poor in emerging countries. Although many construction projects are going on to expand the network, the total supply of road facilities does not meet the increasing demand. To evaluate such supply of road facilities, an index of road density, which is defined as the total road length divided by the target area, is used. Road density of emerging countries is relatively low, which means that their road network is generally sparse (Central Intelligence Agency, n.d.).

The shape of the network is also important. In emerging countries, individual links are not well connected to each other but many minor streets are branched off from a single major arterial street, typically. In addition, due to the limited amount of road supply, the roles of individual links in the network are not clear. As a result, all traffic including long-distance intercity trips, short-distance local trips, and roadside access trips have to come out to the major street, and it gets congested easily. A well-organized network has nodes or intersections at appropriate intervals between different links, thereby providing multiple routes for road users, preventing local concentration, and distributing traffic. In addition, a network with hierarchical structure provides different travel speeds by different levels, which let road users select different routes depending on their trip length.

When we consider individual links, road width or number of lanes are usually not sufficient. In emerging economies, roads with multiple lanes in each direction are often limited to urban areas, while the rest of the country is covered by only two-lane two-way roads. If they are not sufficient to meet traffic demand, congestion will occur. Once congestion occurs, the excess demand accumulates upstream and may extend beyond the link. Then congestion spreads to other links connected to that link, even if their capacity is sufficient. This degrades the functionality of the entire network. Therefore, in order to counter congestion, it is important at first to identify the point at which congestion originates, which is called a bottleneck.

In urban areas, bottlenecks are almost always at intersections. Intersections are the nodes of multiple links, where traffic from different directions intersect, and thus the right of way must be divided among them. Therefore, the time available for each approach is inevitably more limited than that in a simple straight section, resulting in smaller traffic capacity causing more congestion. This is the same for both unsignalized and signalized intersections. In addition, there is not only straight traffic but also right- and left-turning traffic at

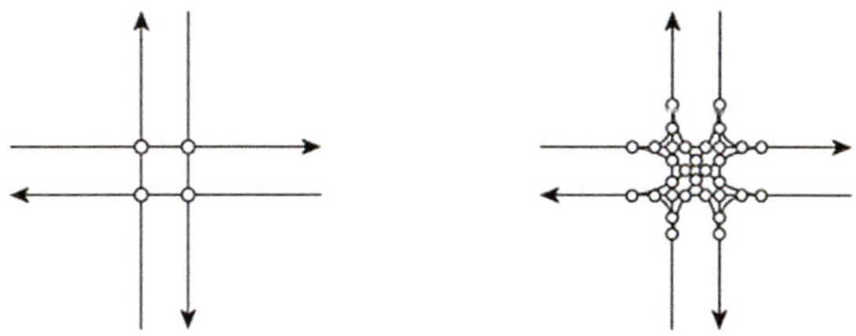

Figure 3.3 Conflict points at an intersection

intersections, resulting in more complex movements. As Figure 3.3 shows, the number of conflict points at a cross intersection with only straight traffic is 4, but when right- and left-turning traffic is added, the total number of conflict points becomes 32. As a result, traffic capacity is even more reduced because it is necessary to take a margin to avoid collisions at these conflict points.

When traffic signals are installed at intersections, congestion can occur due to improper signal settings. Specifically, the signal allocates the right of way (green time) to two or more different directions, but if its ratio is not consistent with the ratio of traffic demand in each direction, congestion will occur from the direction with insufficient green time. Since demand varies with time, the ratio of green time should be varied accordingly, but this requires sensing information using vehicle detectors, which is often difficult in emerging countries.

Traffic capacity also decreases in simple straight sections due to various factors. Capacity may be reduced not only when the width or number of lanes is physically reduced, as in the case of roadwork, but also in other cases. Road surface conditions are also possible causes of congestion. Many roads in emerging countries are unpaved or have a lot of pit holes even paved, where vehicles have to slow down, that may lead to traffic congestion. Rainfall is also known to be a factor to reduce traffic capacity, even in developed countries. In tropical countries, heavy rainfall often occurs in a short period of time during the rainy season. If the roads in those countries do not have adequate drainage system, heavy rains can flood the roads, reducing the space available for traffic and allowing vehicles to travel only at low speeds. This is another factor contributing to reduced traffic capacity.

In addition to the capacity loss caused by road structures and facilities described earlier, there is also capacity loss caused by road users. In emerging countries, road users often do not have a high level of normative awareness, such as their driving manners, typically. As a result, selfish behaviors of some people may affect traffic flow and exaggerate traffic congestion. A typical example is disordered on-street parking. Inappropriate on-street parking on busy arterial roads and other locations has a big impact on traffic flow, which reduces the effective road width available for passing vehicles and lowers traffic capacity. The same can be said for buses that stop inside intersections to have passengers to board and alight, as is common in emerging economies. As mentioned earlier, intersections have lower traffic capacity than simple straight sections, so on-street parking in such location has a more negative impact. There are more cases where traffic capacity is reduced by selfish drivers who do not follow traffic norms, such as making aggressive left-right turns, merging, and driving without observing lanes.

Pedestrians also become a factor of traffic capacity reduction. The presence of pedestrians protruding from the sidewalk into the roadway leads to a reduction in effective road width. However, this is often caused by parked vehicles on the sidewalk. Pedestrians crossing outside of designated pedestrian crosswalks (or jaywalking) also reduces traffic capacity by making vehicles slow down (Liu, G. et al., 2018). In emerging countries, many pedestrians are seen crossing even where pedestrian bridges are provided and at-grade crossing is prohibited, causing traffic congestion as well as traffic accidents.

Street vendors operating on the road are another factor influencing road users. There are many street vendors (of food, beverages, and goods) in downtown areas in emerging countries that occupy the curbside to conduct their business. Similar to on-street parking, these businesses reduce the effective width of the carriageway, thereby decreasing traffic capacity. In addition, gathering pedestrians and vehicles stall around them, further reducing traffic capacity. In addition, these street vendors prefer to select locations that can attract more customers, which are often near intersections, and thus have a greater impact on traffic (Getu, N. et al., 2024).

As described earlier, it can be said that not only road infrastructure is poor in emerging economies, but also the characteristic behaviors of various road users affect both the traffic demand side and the traffic capacity side of the road, which in turn exacerbates traffic congestion.

3.5 TRAFFIC REGULATION, CONTROL, AND MANAGEMENT

Shinji Tanaka

Proper traffic management is essential to reduce the problems of traffic accidents and traffic congestion described so far. Traffic management is the general term for the actions taken by road administrators or traffic administrator on traffic flow, and can be categorized as follows according to its content:

1. Traffic rules
2. Traffic regulation
3. Traffic control
4. Enforcement
5. Traffic demand management

As mentioned in the previous section, road users in emerging countries are often less disciplined in their driving manners and other traffic behaviors than in developed countries, which contributes to traffic accidents and traffic congestion. Therefore, maintaining them through proper traffic management is very important to achieve safe and smooth traffic. Each of these details is described in the following sections.

3.5.1 Traffic Rules

Traffic rules are the basic rules that road users must follow to ensure the safety and smoothness of road traffic, and violations of these rules are punishable by penalties. In emerging countries, however, there are cases where these basic rules are not always observed. Some of the major examples are described in the following subsections.

3.5.1.1 Traffic Segmentation

There is a rule of the road to pass on the right-hand side or left-hand side in every country. This allows traffic facing oncoming traffic to pass each other without collision. If roadways and sidewalks are separated in urban areas, vehicles are required to pass on the roadway and pedestrians on the sidewalk to ensure safe and smooth traffic in urban areas where many road users exist. These are basic traffic rules used throughout the world.

In emerging countries, however, there are many cases where these basic rules are not followed. For example, some vehicles run in the reverse direction on roads where a median strip separates inbound and outbound traffic because they do not want to increase the travel distance. Some motorcyclists ride from the crowded roadway onto the sidewalk. Since these are dangerous behaviors that can lead to traffic accidents, it is necessary to change society's perception of such behaviors as well as penalize them through enforcement.

3.5.1.2 Driving License

Driving a car requires specialized knowledge, and if driven incorrectly, it can cause serious damage to surrounding people and objects. A driving license proves that the driver has the knowledge and skills necessary to drive a car, and the driver is required to carry it when driving a car on public roads where there are many general public. Driving license systems vary from country to country, and the ease of obtaining a license and the awareness of compliance with the rules vary as well.

In emerging countries, the ease of obtaining a driving license may make it simpler to obtain a license, but this also means that there is less awareness of the need to comply with rules and regulations, and unlicensed driving is more likely to occur. This is even more likely to happen with small motorcycles than with four-wheeled vehicles, and it is sometimes found that boys and girls are driving small motorcycles who are obviously under the age for a license. This is probably because the adults around them allow them to do so, but considering the consequences of an accident, society as a whole must work to foster an awareness to comply with the licensing system.

3.5.1.3 Prohibition of Drunk Driving

Since alcohol consumption affects each phase of driving – perception, judgment, and maneuver – and can cause delays or errors in these actions, driving a vehicle under the influence of alcohol is prohibited in most countries. However, awareness of compliance varies from country to country, and this is the same in both developed and emerging countries.

In emerging countries, many people drink alcohol during traditional events such as New Year's holidays and festivals, which often lead to driving under the influence of alcohol in such an uplifted atmosphere. The frequency of traffic accidents also tends to increase during these times, which has become a major problem. In order to improve this situation, it is necessary not only to strengthen penalties for drunk driving, but also to publicize and educate people about the significant impact of loss caused by drunk driving.

3.5.1.4 Protection of Pedestrians

Since pedestrians are the most vulnerable of all road users, vehicles such as bicycles and motorized cars are required to make every effort not to obstruct pedestrian's walking and to protect them. Vehicles must give way to pedestrians at crosswalks when pedestrians try to cross the road even if there are no traffic signals.

However, in many emerging economies, vehicles with larger bodies and higher speeds seem to have priority over pedestrians, and pedestrians cross the roadway in between them. This is a significant traffic safety problem. Thus, on the one hand, education, awareness, and enforcement of pedestrian priority must be promoted. On the other hand, pedestrians should also understand that it is important to cross the road at designated crosswalks, and that crossing in a disorderly manner, as described in the previous section, is a behavior that puts themselves in danger.

3.5.1.5 Passenger Capacity and Maximum Loading Capacity

If passenger cars and buses carry more passengers than their capacity, or cargo vehicles carry more cargo than their maximum payload, it can affect the running stability of the vehicle and lead to traffic accidents. Once an accident occurs, more passengers and cargo will be damaged. In addition, insufficient acceleration from a stopped status or on an uphill slope can affect following vehicles, causing congestion and traffic jams. Excessive loads that exceed the vehicle's performance place a heavy load on the engine, which also increases CO_2 and other emission gases. In addition, it is known that the impact on road pavement is generally proportional to the fourth power of the load, causing ruts, cracks, and sinking of the pavement, which in turn affects road infrastructure facilities. To prevent such impacts, it is very important to comply with the passenger capacity and maximum loading capacity specified for each vehicle.

In emerging countries, there are many cases of large commercial vehicles, especially buses and trucks, that are over-capacity and overloading. It poses a major challenge to road safety and road infrastructure maintenance. These are typical problems of external diseconomies, in which the drivers gain economic benefits in exchange for inconvenience to others; therefore, monitoring and control from the surroundings is essential.

3.5.2 Traffic Regulation

Traffic regulations are designed to regulate traffic by restricting some of the behaviors of road users at specific times and places, thereby ensuring the safety and smoothness of traffic as a whole. The content of the regulation is presented to road users by means of signs and road surface markings. Violators are subject to penalties, as non-compliance can have a negative impact on traffic. Typical traffic regulations that are considered particularly important in emerging countries are described in the following subsections.

3.5.2.1 Maximum Speed Limit

Speeding is one of the main causes of traffic accidents. In particular, vehicle speed has a significant impact on the occurrence of so-called serious accidents involving fatalities or serious injuries. As vehicle speed increases, the likelihood of an accident increases as the time between the driver, pedestrian, or other subject recognizing a hazard and making a decision or taking action to avoid the hazard becomes shortened. In addition, since kinetic energy, the impact of a collision, is proportional to the square of the speed, faster speed collisions are more likely to result in more serious accidents. To reduce these risks, it is necessary to establish appropriate maximum speed regulations according to the road environment and traffic conditions.

It is mentioned in Section 3.1.3 that in emerging countries, automobiles tend to be treated as a social status more than a means of transportation. In some cases, the performance of a vehicle and the skill of the driver who operates it, that is, how fast the vehicle can run, are similar elements of self-expression to others. Such vehicles traveling at high speeds ignoring the surrounding road environment and traffic conditions are a major risk to traffic safety, and this must be properly controlled through speed limit regulations.

3.5.2.2 No Overtaking Regulations

On a two-lane two-way road where inbound and outbound traffic are not separated by a median, overtaking behavior using the oncoming lane poses significant traffic safety risks, such as head-on collisions with oncoming vehicles and roadway departure crashes due to overspeed. For safe overtaking, the required overtaking sight distance must be maintained based on the speeds and positions of the overtaking vehicle, the overtaken vehicle, and the oncoming vehicle. However, this may be difficult to maintain in areas with poor visibility such as curved sections or near the top of an uphill grade, or in sections with many oncoming vehicles. Overtaking is prohibited in such areas to reduce the risk of traffic accidents.

In emerging countries, road structures with separated inbound and outbound lanes are rare except in urban areas, and two-lane two-way roads often cover most of the country. In some cases, lane boundaries, center lines, and outside lines are not clearly marked, and shoulders are inadequately paved, depending on the state of road maintenance. Dangerous driving behaviors peculiar to emerging countries can also be seen when overtaking, such as double overtaking, that is, a vehicle trying to overtake another overtaking vehicle from behind, and overtaking using the shoulder or outside the carriageway, which can lead to traffic accidents. Therefore, the risk of traffic accidents caused by overtaking behavior is higher in emerging countries than in developed countries, and this risk must be properly controlled through no overtaking regulations.

3.5.2.3 On-Street Parking Regulations

Trips are taken to perform some activities at a destination in most cases, so the purpose of travel cannot be achieved if the used vehicle cannot be parked. On-street parking is done when there is no parking available at the destination, when parking is not available because the parking lots are full, or when parking is available but it is more convenient to park on the street. On-street parking occupies a certain space of the road, which reduces the space for running vehicles, causing road congestion and traffic jams. In addition, on-street parked vehicles are an obstacle on the road for moving vehicles, causing accidents such as collisions or missing pedestrians who appear behind the parked vehicles. On-street parking regulations prevent these risks. Regulating on-street parking in downtown area also makes it more difficult for people to come there by car, thereby reducing the traffic demand attracted to urban centers and easing road congestion.

On-street parking regulations are needed primarily in urban areas with high traffic demand. Types of regulations include no stopping, no parking, parking time restrictions, parking restrictions by vehicle type, and parking charges. These should be selected in consideration of the role of the road (i.e., whether the road should serve through traffic or access to the district), the width of the road, the amount of through traffic, traffic demand in the district, and the supply of off-street parking spaces in the vicinity. Flexibility to change regulations based on traffic conditions, such as day of the week and time of day, is also important for effective traffic management.

There are few emerging countries that have strict on-street parking regulations, and in many cases, the curbside lanes are completely occupied by on-street parking in commercial areas in urban centers, as shown in Figure 3.4. In order to solve problems such as traffic congestion, it is essential to manage parking in urban centers, including appropriate on-street parking regulations. Double on-street parking, parking in inappropriate locations such as inside intersections and on sidewalks, is also common, and its impact on traffic flow is so great that it needs to be eliminated through enforcement.

Figure 3.4 On-Street Parking

Source: (Photo taken by Shinji Tanaka)

3.5.3 Traffic Control

Traffic control uses specific control devices to control traffic flow. A typical example is traffic signals at intersections. At intersections where traffic in different directions intersects, green lights are given to traffic in a specific direction at timed intervals to regulate the traffic flow and ensure traffic safety. The orderly flow of traffic also contributes to smoothness and allows for higher traffic capacity than at intersections without signals. As mentioned earlier, in emerging economies where user discipline of traffic is less well established, users would not be able to judge priority relationships if there were no traffic signals, and intersections could easily become clogged with incoming vehicles from all directions. It can be said a traffic signal that can unify the users' intentions is highly useful in emerging countries with a diverse mix of road users.

In addition, hand signals and traffic guidance by traffic police officers and traffic control personnel can also be considered a type of traffic control. Such manual traffic control is widely used in emerging countries, and is an effective method of control until facilities such as traffic signals are installed. However, because the timing of switching right of way is left to the subjective judgment of the person in charge, the control cycle, which corresponds to the cycle length of a traffic light, generally tends to be long, with some cycles exceeding 300 seconds (= 5 minutes) in some places. As the cycle becomes longer, the saturation traffic flow rate on the traveling side gradually decreases, while the delay time on the stopped side increases, resulting in inefficient control in many cases. To avoid this, it is desirable to control with a cycle time of 150 to 180 seconds even during congested periods when traffic demand is high.

In some countries, even though traffic signals have been installed with development assistance, they are not used (not turned on) and are controlled by police officers using hand signals. This is because the electric power supply may not be available due to power outages, maintenance and repairs may not be carried out in case of equipment failure, and police officers may not trust traffic signal control or prefer to control traffic by their own hands rather than using traffic signals. In such a case, it has become a place for police officers to show off their power.

3.5.4 Enforcement

The traffic rules, traffic regulations, and traffic controls discussed earlier cannot be effective unless all road users comply with them. If there is no enforcement against those who do not follow the rules, the number of those who do not follow the same rules will gradually increase, and the roads will eventually fall into chaos. As a result, the safety and smoothness of road traffic will be compromised. Therefore, proper enforcement of violators is essential to make traffic rules, traffic regulations, and traffic control effective.

As enforcement requires coercive power against violators, it is carried out by traffic administrators, which in many countries are the police. The most common form of enforcement is manual enforcement by police officers and traffic administration officials, but enforcement using some equipment such as CCTV cameras and speed-measuring radar is also widely implemented.

In emerging countries, some violators may bribe police officers in charge of enforcement to get traffic violations overlooked, or conversely, police officers may demand bribes from violators. These are not problems that can be solved in a short period of time, as they are greatly influenced by the normative consciousness of society as a whole. However, since these overlooked violations will continue to have a negative impact on society, efforts to reduce such corruption through steady education and awareness-raising activities are necessary.

3.5.5 Traffic Demand Management

Methods to reduce traffic demand on a road network to prevent traffic congestion are called traffic demand management. This can be done by encouraging people to change their routes from congested ones, to change the time of travel, and to change the means of transportation, or by discouraging the use of the road itself. One of these methods of regulating the use of vehicles is the license plate regulation. This is to designate the days when road users can use their vehicles according to the number on their license plate, specifically whether the first digit on the plate is even or odd, or whether the number is 0–9.

Since license plate regulations do not require special equipment or facilities for implementation, they are relatively easy to introduce even in emerging countries. However, some households take measures such as owning multiple vehicles, and it has been pointed out that the effect may not be as great as expected in such cases.

The traffic management described earlier will be more effective if they are combined and implemented in an integrated manner, for example, by implementing traffic control in conjunction with traffic regulation, rather than by implementing each of them separately. And in order for these to function effectively, enforcement is particularly important in emerging countries.

3.6 FREIGHT TRANSPORTATION AND LOGISTICS

Takayoshi Futose

3.6.1 Introduction

Emerging countries are in a dynamic transition of the economy from primary, light industry or informal service sectors to heavy industry or formal sectors. Through the transition, freight transportation plays a pivotal role in the local, national, and international economy.

There are various freight transportation modes, including truck, ship, inland water, air, rail, and pipelines, and the modal share varies depending on geographical conditions, economic situation, and the level of infrastructure development. Despite encouragement to shift to more environmentally sustainable modes, road transport remains the dominant mode in many regions of the world and plays an important role in lower-income countries. Especially for landlocked countries, road transport is often the only available mode to access global markets.

Major freight-related traffic generators include industrial estates, commercial centers, traditional markets, ports, inland container depots, truck terminals, freight rail stations, and airports and could cause a huge traffic impact. Traffic congestion could affect the smooth and reliable goods movement which contributes to economic development. Additionally, freight transportation contributes to air pollution, noise emissions – impairing road safety – and burden on road infrastructure.

Road-based freight transport industry allows individuals to become entrepreneurs and enter the business by owning their vehicles. It could be considered as important to give opportunities for entrepreneurship and job creation, especially in developing and emerging countries. As a result, this sector is highly fragmented and individualized in terms of ownership and operation. Most of the trucks are owned by small operators with only one or a few trucks, and there are a small number of large operators with many trucks. In this condition, how to improve operation efficiency is one of the major challenges, and it contributes to minimize the negative impacts of freight traffic.

Management measures to optimize the freight movement and mitigate the negative impacts are necessary to be implemented, and ICT is expected to be utilized in the field.

3.6.2 Problems of Road-Based Freight Transportation

Various traffic problems can be induced by road-based freight traffic. Increased traffic volume, partly due to the inefficient vehicle operations in carrying goods, causes severe traffic congestion (GIZ, 2010). The space for handling goods is limited, and trucks have no choice but to park on the roadside, reducing the capacity for other vehicles and pedestrians. Due to the mixture of heavy vehicles with smaller vehicles and non-motorized vehicles, over-usage of old vehicles, and extended working hours of drivers, the risk of traffic accidents is increased.

Some directly affect the quality of life of the people, and others contribute to global criminal change. Green house gas (GHG) emissions and local air pollution must be minimized by using cleaner vehicles or modes and through optimization of the logistic operation. Traffic noise has severe impacts on health and overall QOL for people, especially those who are in the vicinity of residential areas. Land use must be regulated to separate freight traffic generators and residential areas.

Freight traffic may affect also the urban economy. Heavy freight vehicles cause accelerated deterioration of roads and bridges, especially if the maximum permitted vehicle axle loads and weight are ignored. Without proper enforcement, it could cause huge economic losses. And increasing operating costs of logistics are competitive disadvantages to sustain the economic activities of the city and attractiveness for the investors.

3.6.3 Options for Countermeasures of the Problems

In this part, various countermeasures which are planned and practically implemented in emerging countries are introduced.

3.6.3.1 Institutional Measures

Legislation and regulation on the national or regional level may be the prerequisites for implementing the measures to handle freight traffic.

Adequate legal framework on the national or provincial level is one of the first prerequisites for policy implementation on the local level. National or regional legislation allows for more local governments to impose their management measures.

The fee structure for vehicle licensing and the taxation policies may influence the development of the industrial structure. Taxation could not only cover the social costs and environmental damages but also optimize the traffic. And taxation for high-emission vehicles gives preference to lower-emission vehicles. Introducing the right policies can benefit in terms of air quality, climate, and economy if advanced second-hand vehicles could be imported reasonably. Global shift into electric vehicles (EV) including in emerging countries has been supported by the selective taxation scheme.

To ensure the safety and service quality, introducing business licensing with standards that operators shall meet is necessary. Emission requirements or standards for vehicles being imported or for new registration contribute to promote the renewal into more clean and fuel-efficient vehicles. Regular vehicle inspection to check and enforce whether vehicles meet the prescribed emission requirement is necessary to ensure the effectiveness of environmental policies. On the other hand, it shall be noted the licensing process may cause corruption and informal payments.

3.6.3.2 Traffic Management on Freight Traffic

For the bottlenecks or blackspots where freight vehicles contribute to the congestion, measures to manage the traffic flow and transport infrastructure must be conducted.

Separation of freight vehicles and other vehicles providing truck lane contributes to traffic safety along a major corridor with a certain volume of freight traffic. It allows heavy freight vehicles to drive through only on one lane, usually the lane with distance from the sidewalk, to protect pedestrians from the risk of conflict and emissions. Restriction on freight vehicles using a section at certain hours of the day is implemented in many cities and countries. In most cases, freight vehicles are restricted according to the time, vehicle size, and times. Restricting heavy vehicles may encourage the use of smaller vehicles (including passenger cars) with overloading. Figure 3.5 visualizes the access restriction measures in Metro Manila, Philippines, from the ordinance

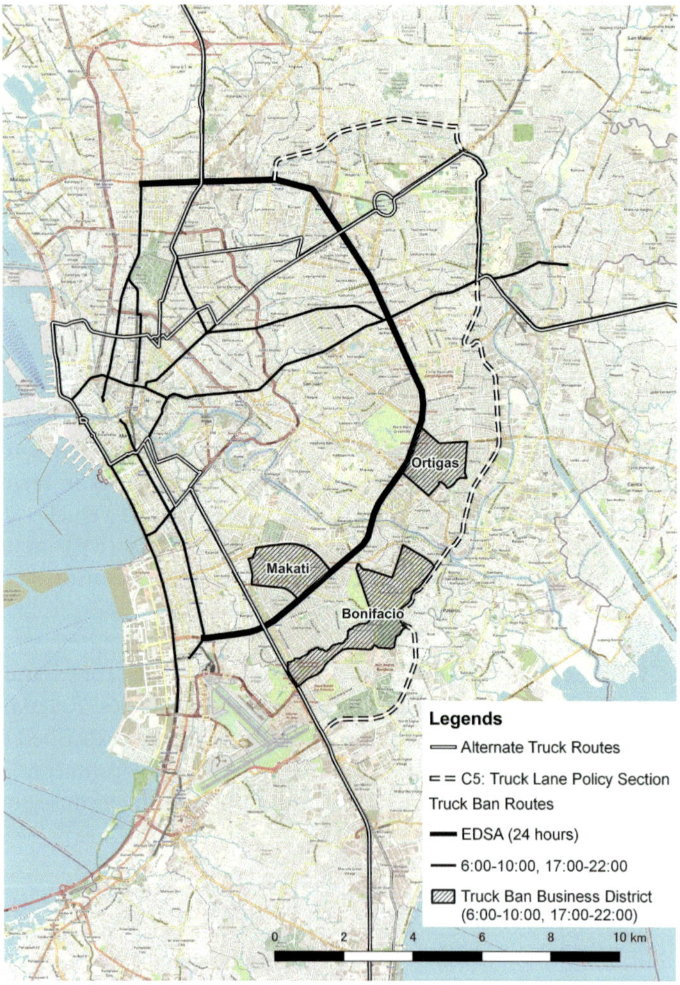

Figure 3.5 Access Restriction of Freight Traffic in Metro Manila, Philippines

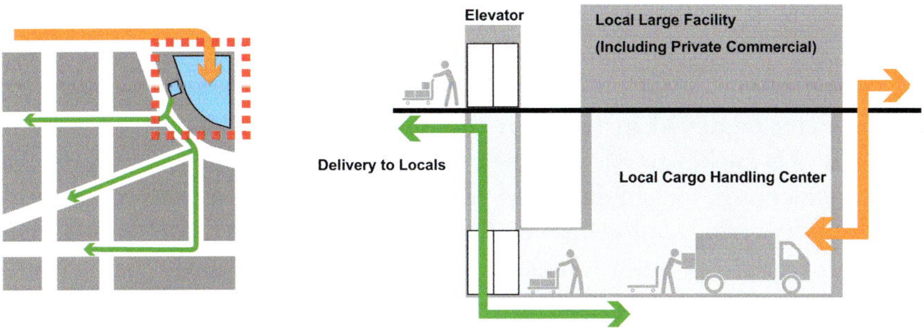

Figure 3.6 Conceptual Image of Local Cargo Handling Center

(Metropolitan Manila Development Authority, 2022). Since the economic stagnation due to COVID-19, the access restriction is temporarily suspended, with the exception of EDSA (Epifanio de los Santos Avenue) (as of April 2024).

Not only are freight vehicles circulated, but stopping for loading and unloading also causes locational traffic congestion. Providing loading and unloading spaces for trucks may significantly contribute to mitigating traffic congestion. How to provide the space has several options including off-street space, sidewalk which is notched out as bay and on-street space (for one-way road or wide road). As an alternative, a local cargo handling center, which can be shared by several operators, is provided in an off-street or underground facility (e.g., Shibuya-Fukuras, n.d.). The conceptual image is shown in Figure 3.6.

3.6.3.3 Urban and Transportation Planning

Not only traffic management measures, but policies on urban and transport plans are also necessary to tackle the long-term issues. Specifically, regulating and guiding the location of the freight traffic generators can mitigate locational congestion and achieve effective urban traffic movements. Many megacities are located along the river, because of the advantage for economic activity through shipping. The vicinity of the CBD or port facilities causes serious traffic problems. Logistic estates shall be developed at the vicinity of interchange, other industrial estates, and traffic nodes. To establish strategies for the improvement and implementation of policies, accurate and comprehensive data should be collected and analyzed. In some countries, a comprehensive survey on the net flow of goods is conducted regularly.

3.6.3.4 Efficiency Improvement by Private Sector

In a lot of countries, most of the truck owners are individual operators and improvement of the operation efficiency is one of the critical issues. For example, vehicles carry more cargo in one direction, but less in another direction. If these directional gaps could be mitigated, the required number of vehicles to carry the same number of cargos would be reduced. For the operation optimization aiming to deliver the same amount of goods with less vehicles deployed in terms of the number of vehicles, vehicle size or vehicle-km is the key to mitigate the problems caused by freight traffic. In the macro-viewpoint, consolidation of freight services is the effective measure. In the short term, cooperative delivery by different operators is expected to improve the operation efficiency. In the middle or long term, reforming the urban logistic industry into consolidated service shall be intended. In the micro-viewpoint, freight operators are expected to optimize the routes to visit dispersed delivery points, in order to improve the amount of cargo delivered per vehicle-hour or vehicle-km.

3.6.4 Technological Intervention

To implement the countermeasures effectively and efficiently, technological innovations could be applied. In emerging countries, leapfrogs from traditional operation to state-of-the-art technology are expected and promoted. However, such technology requires the users who are mainly freight operators to be able to adapt and utilize technology. And it should also be noted that, just introducing technologies without strategy in the local contexts is not able to address the challenges. This section introduces the case studies of the technological intervention in the freight industry in emerging countries:

3.6.4.1 Transport Integrated Management System (Kenya)

Under the Eastern Africa Regional Transport, Trade and Development Facilitation Project by Initiative for the Development of Africa (IDA), the Transport Integrated Management System (TIMS) platform was established in 2016. TIMS is an online platform incorporated with vehicle registration, transfers, licensing, and inspection. TIMS has reduced the time and cost to apply for necessary licensing for all drivers including freight traffic. The database of the vehicles is linked to the government institutions and verifies the information of the person and vehicles (World Bank, 2021).

3.6.4.2 Terminal Appointment Booking System (the Philippines)

Freight vehicles entering the container terminal of Main Ports in Manila should reserve their arrival times in advance via Terminal Appointment Booking System (TABS), which is an online platform developed in partnership with the government, terminal operators, and Australian developer. The reservable number is fixed by time zone and traffic demand is equalized, contributing to mitigation of traffic congestion and improvement of operation efficiency. Fines are imposed for delayed or early arrivals (Asian Terminals Incorporated, n.d.).

3.6.4.3 Quality Control Information

GPS and telematics help to improve operation performances. To detect thefts or traffic accidents, hooking/unhooking of trailers and opening/closing of container doors are reported to the operation managers. Over-speeding and deviations from the supposed route are also detected immediately, to control the traffic behaviors of the drivers. A technology can monitor the temperature inside the containers and fuel amount and visualize them by time-zone. Not only the vehicles, the real-time information of the commodities also contributes to the efficient urban freight traffic movement.

3.6.4.4 Digital Freight Platforms (Vietnam, et al.)

A mobile application visualizes the real-time locations of the goods and freight vehicles and supports the matching for the owners of both. The system looks like a passenger-hailing service. The arrangement fee is automatically estimated from the previous operation records and proposed to the shippers. The drivers find and accept orders considering their travel plans. The service is beneficial for the stakeholders, and the number of vehicles to carry the cargo could be reduced (Logivan, n.d.).

3.6.4.5 Drone Delivery (Rwanda, et al.)

Delivery service by drone is an emerging service that is considered beneficial in terms of speed, flexibility, environment, and the reduction of operational and overhead costs (GIZ, 2021). In Rwanda, drones are used to make emergency deliveries, including blood for transfusions or other medical items.

3.6.5 Summary and Future Direction

Freight traffic is a critical component of economic development in the city and causes serious traffic problems. Addressing these traffic problems requires integrated measures including legal framework, demand management, operation optimization, and land use policies. Technological solutions also present opportunities to modernize and streamline freight operations. In emerging countries, technologies are expected to enable a leap from traditional freight operations and have been actively promoted, although challenges remain in their adoption and utilization.

3.7 CONCLUSION

Fumihiko Nakamura

In this chapter, we have focused on the evolution of transportation policies, traffic accidents, motorization, and congestion in emerging cities with respect to cars, motorcycles, and freight.

The transition of transportation in cities has reached the age of the automobile through a historical process. In the past, walking, horse-drawn carriages, bicycles, and motorcycles were the main modes of transportation, but as cities developed and economic growth progressed, they were replaced by the automobile. This transition is particularly pronounced in emerging markets.

As cities expand, transportation moves from human power (walking) to mechanical or motorized power (bicycles, motorcycles, and eventually cars). Geographical, economic, and trade conditions also have a significant impact on this change. For example, bicycles are more popular in flat areas, while motorcycles and cars are more popular in mountainous areas. In economically affluent areas, the transition from bicycles and motorcycles to cars is more rapid, but in low-income areas, the transition may be delayed.

Because owning a car is seen as a symbol of social status, more urban residents buy and use cars as incomes rise. Once people own a car, they tend to use it more frequently. However, in areas where cities are densely packed, the physical constraints of roads make it difficult to completely replace cars. For example, in cities like Hanoi, where motorcycles were the main mode of transportation, the urban design and transportation infrastructure were built to accommodate this characteristic, so while there was a smooth transition from bicycles to motorcycles, it is difficult to make a complete transition from motorcycles to cars.

The transition from bicycles and motorcycles to cars is not just a technical change, but a complex issue involving urban planning, traffic space, and parking issues. Efficient management of public transportation and road space is important to address these issues.

Motorcycles are popular in many Asian countries, and the timing of their popularity varies from country to country. Because motorcycles are two-wheeled vehicles, they are less stable than four- or three-wheeled vehicles, and traffic accidents involving motorcycles are more common. In countries where motorcycles became popular particularly quickly, this safety problem is particularly serious, and the founder of Honda Motor Company, a famous motorcycle manufacturer, established the IATSS with his own funds with an aim to reduce traffic accidents.

Motorcycle safety measures mainly consist of the 3 E's (Engineering, Education, and Enforcement). Engineering requires improvements to roads and traffic management. For example, the construction of dedicated lanes and traffic lights for motorcycles is being promoted. In terms of education, road safety education is lacking in many countries because of the ease of obtaining a driving license. Therefore, there is a need for more education on traffic laws and safe driving. Finally, in terms of enforcement, if the ability to enforce the law is insufficient, there is a limit to how much accidents can be reduced.

Developed countries, including Japan, have well-established driver licensing systems and traffic safety education, but these are often lacking in emerging countries, and there are still challenges to reducing accidents. Initiatives such as Honda's motorcycle training center in Thailand have had some success, but more needs to be done overall.

Motorization refers to the increasing prevalence of cars. This phenomenon is particularly pronounced in emerging markets, where the number of vehicles on the road increases as the economy grows and roads improve. As roads improve, motorization accelerates, but the lack of sufficient road infrastructure in emerging markets can lead to congestion and accidents.

In addition, as motorization progresses, traffic volumes increase rapidly, making it difficult to move between and within cities, and the size of vehicles also increases. As vehicles become larger, road widths and the number of lanes required also increase, causing congestion as road capacity cannot keep up.

In urban planning, to cope with motorization, it is necessary not only to expand the road network, but also to develop a multi-layered network to disperse traffic and implement policies to control traffic demand. It is also important to develop public transportation, which can help reduce road congestion.

Traffic congestion is caused by two main factors: the demand side and the supply side. On the demand side, economic growth leads to an increase in the number of vehicles, resulting in a greater variety of vehicles on the road. In emerging economies, the number of vehicles is increasing rapidly due to economic growth, and a wide variety of vehicles, such as luxury cars, old cars, and overloaded buses, are sharing the same roads, which complicates vehicle speeds and traffic flow, and leads to congestion. In addition, as vehicles become larger, this can exacerbate congestion.

On the supply side, one of the causes is inadequate road infrastructure. In emerging markets, the road network is not well developed, and there is a lack of road capacity, especially in urban areas, which leads to traffic congestion. Due to the shape and density of the roads, as well as the poor connectivity of the network, the flow of traffic is not dispersed and vehicles are concentrated on the main roads.

To improve traffic congestion, it is important to improve the road infrastructure and design the network to encourage the dispersion of traffic flows. In urban areas, it is also necessary to identify bottlenecks and eliminate the factors that cause congestion.

As described earlier, we have explained the transition of bicycles, motorcycles, and cars as modes of transportation and their impact on road safety, the development of motorization, and the causes of traffic congestion. In the development of cities, it is important to improve transportation infrastructure, as well as to adjust transportation demand and improve public transportation.

REFERENCES

Asian Terminals Incorporated. (n.d.). *Terminal appointment booking system: User's guide*. Retrieved from https://www.asianterminals.com.ph/uploads/ATI_TABS_UsersGuide_download.pdf

Central Intelligence Agency. (n.d.). *The world Factbook: Roadways*. Retrieved from https://www.cia.gov/the-world-factbook/field/roadways/country-comparison/

Deutsche Gesellschaft für Internationale Zusammenarbeit (GIZ) GmbH. (2010). Urban freight in developing cities. In B. O. Herzog (Ed.), *Sustainable urban transport project (SUTP) sourcebook module 1g*. Retrieved from https://sutp.org/publications/urban-freight-in-developing-cities-2/

Deutsche Gesellschaft für Internationale Zusammenarbeit (GIZ) GmbH. (2021). Delivering sustainability: Urban freight in developing cities. In *Sustainable urban transport project (SUTP) sourcebook module 1g* (2nd ed.). Retrieved from https://sutp.org/publications/delivering-sustainability-urban-freight-in-developing-cities/

Getu, N., Kifle, D., Mesfin, A., Yifru, W., Tamene, M., & Sewunet, A. (2024). Analysis of street vendor effects on urban arterial road. *Transportation in Developing Economies*, *10*(1), Article 1. doi:10.1007/s40890-023-00188-5

International Association of Traffic and Safety Sciences (IATSS). (2015). *Traffic and safety sciences: Interdisciplinary wisdom of IATSS*. Japan Times Ltd.

Liu, G., Lu, J., Wang, J., Jing, X., He, J., Gao, P.,. . . . & Li, R. (2018). Influence of pedestrians' jaywalking behavior on vehicle flow. *International Journal of Modern Physics B*, *32*(22), 1850233. doi:10.1142/S0217979218502330

Logivan. (n.d.). *Digitising and optimising freight transportation in Vietnam*. Retrieved from https://logivan.com/

Lu, M., Sun, C., & Zheng, S. (2017). Congestion and pollution consequences of driving-to-school trips: A case study in Beijing. *Transportation Research Part D: Transport and Environment*, *50*, 280–291. doi:10.1016/j.trd.2016.11.029

Metropolitan Manila Development Authority. (2022, June 8). *Truck Ban Ordinance*. Retrieved from https://mmda.gov.ph/home/20-faq/5564-truck-ban-ordinance-updated-faq-june-8–2022.html

Ramanayya, T. V. (1988). Highway capacity under mixed traffic conditions. *Traffic Engineering & Control*, *29*(5), 284–287.

Shibuya-Fukuras. (n.d.). *Local Cargo handling center ESSA*. Retrieved from https://www.shibuya-fukuras.jp/essa/

World Bank. (2021, July 7). *Transforming Kenya's transport sector through digital integration*. Retrieved from https://www.worldbank.org/en/news/feature/2021/07/07/transforming-kenya-s-transport-sector-through-digital-integration

Chapter 4

Paratransit in Emerging Countries

Toshiyuki Okamura

4.1 WHAT IS PARATRANSIT?

4.1.1 Introduction

In emerging countries, less organized mobility services, provided by the small-scale or sole proprietors, play a very important role in urban transportation. They have emerged spontaneously in almost all low-income and lower-middle-income countries under locally specific conditions, and have developed independently in response to people's demands. Those modes are collectively referred to as paratransit (Behrens et al., 2023). There are two main types of service; route-fixed collective transit (equivalent or close to bus service), and for-hire transit (equivalent or close to cab service) that provide door-to-door individual transportation. The characteristics of this paratransit mode are that it is spontaneous, market-responsive, and not necessarily operated under a formal system (informality) and that it tends to be a small-scale or sole proprietorship business that uses small- to medium-sized vehicles with low initial costs. Furthermore, these two characteristics tend to reveal a lower level of technology, level of service, and working environment than is common in low- and lower-middle-income countries, while flexible and market-responsive services are provided, resulting in a large number of users.

Despite the fact that paratransit is extremely diverse from country to country and city to city and has undergone its own development, it shares many of the common characteristics described earlier. To understand paratransit in emerging countries, it is necessary to focus on this diversity as well as its universality when viewed from a global perspective. In this chapter, this type of mode is broadly considered as "paratransit in emerging countries," and its current status, challenges, and future prospects are discussed.

DOI: 10.1201/9781003467403-5

4.1.2 The Terms Paratransit and Informal Transportation

The term "paratransit" is one of the most commonly used terms to describe the transportation modes described in Section 4.1.1 in emerging countries (in developed North America, the term "paratransit" refers to demand-responsive transportation modes that primarily serve the disabled and elderly, which is not the subject of this chapter). In addition, an equally commonly used term is "informal transportation" or "informal transport" (Cervero and Golub, 2007; Baffi and Lannes, 2021; International Association of Public Transport (UITP), 2024; United Nations Development Programme (UNDP), 2024). In the original meaning of the term, the former focuses primarily on vehicle characteristics, while the latter focuses primarily on institutional characteristics, such as operations, organization, and governance. Nevertheless, the two terms refer to almost the same thing consequently. The scope of these terms is not necessarily rigid, and there is considerable latitude in their use by transportation planning professionals.

In terms of vehicles, the term "paratransit" is commonly used in academic and practical terms to refer to small- to medium-sized vehicles with seating capacity ranging from two passengers (including the driver) to 30 or even 50 passengers. Motorcycle taxies with a seating capacity of two, tricycles (three-wheelers) with a seating capacity of three, and vans with a seating capacity of 15 are easily understood as paratransit, and in fact, many types of transportation are seen in emerging countries as a major player. Then how essential is the difference between bus service with medium-sized vehicles of about 30 or more passengers and large vehicles of about 50 or more passengers and bus (collective service) service with vehicles of lesser size? Certainly, the capacities per vehicle are different, and large vehicles that allow standing seats have relatively high space efficiency in transportation. Also, for vehicles smaller than about 50 passengers, the service availability in relation to road infrastructure is naturally different, since service can be provided on narrower roads. However, there is no threshold that clearly distinguishes between the two, and there is no meaning beyond that of a continuous change in characteristics due to an increase in vehicle capacity. Indeed, (large) buses and paratransit sound distinctly different as terminology. The Bangkok Mass Transit Authority (BMTA), the public bus system in Bangkok, Vientiane (Laos), Phnom Penh (Cambodia), and Yangon (Myanmar), for example, is one publicly owned entity. It certainly seems desirable to treat them as a mass transit system, which is different from paratransit, in terms of transportation planning. However, even such public buses have some flexibility in their operation patterns and drivers' work schedules (e.g., schedules are not necessarily fixed). Furthermore, in the case of small- and medium-sized licensed or franchised bus operators in the private sector, they are more informal in terms of employment status and service level, and have a paratransit character. Therefore, there is no clear and essential difference between (large) buses and paratransit, and in many cases, there is little academic or practical meaning to the different categorization distinctions. The explanation, then, is that paratransit in emerging countries is represented by small- and medium-sized vehicles as the "more typical and more popular vehicle type" size.

An essential characteristic of paratransit in emerging countries is their informality. The essential characteristics of paratransit in emerging countries are that it offers flexible, market-responsive services and, like many types of businesses in emerging countries, has some informality. These characteristics have a variety of externalities for citizens, societies, and cities. In this regard, the term informal transportation has long been generally agreed to be a straightforward term that encompasses many of these characteristics. Antonyms for "informal transportation" include "formal transportation" and "institutional transportation." However, it isn't easy to clarify whether paratransit in emerging countries is informal or not, since there is a wide range and type of informality (Rekhviashvili et al., 2022). The term "informal transportation" can be understood as underestimating the diversity of informality and the positive effects it can have by clearly dichotomizing informal and formal. It can also be said to implicitly mean that the goal of the policy is to transition the current informal transportation to "formal" ones. In fact, integration based on strong government involvement is neither the only approach nor the ultimate goal for reform and innovation of paratransit in emerging countries. Furthermore, for-hire type shared mobility is established under platform apps and is truly informal in the sense that it is not subject to government planning and regulation. On the one hand, it is more formal in terms of services and technology and is quite different from traditional informal transportation. On the other hand, in the sense that it is operated by private proprietors with small, privately owned vehicles, it has similarities with the traditional paratransit.

The term paratransit should not be strictly limited by vehicle size, and the term informal transportation can be considered a limiting term in the sense that it evokes a clear dichotomy between formal and informal. Informality is a widespread phenomenon in emerging countries, and it is not desirable to distinguish it from formality in particular, nor is it desirable to take an overly negative view of informality. In this chapter, therefore, we use the term "paratransit" to refer to public transportation in various emerging countries that have some form of informality. In this broad sense, paratransit is characterized by flexibility and informality of operation, organization, and regulation, which brings with it a variety of positive and negative externalities and can be classified based on various technical, institutional, and social (in)formality. These are discussed in Section 4.2 and beyond.

4.1.3 Role and Negative Externalities of Paratransit in Emerging Countries

This section summarizes the role and (mainly negative) externalities of paratransit in emerging countries, which are characterized by flexibility and informality (Cervero and Golub, 2007; Baffi and Lannes, 2021).

4.1.3.1 Essential Mobility in Emerging Countries

Not all means of transportation for people within a city or region are provided by governments or large enterprises. Even when so-called formal public transportation systems such as metros, BRT, and other mass transit systems exist, few cities in emerging countries have all their mobility needs met by them. Demand-responsive paratransit provides business-oriented transportation services where demand exists. There is no doubt that paratransit is an important transportation option for low-income residents. However, the view that the role of paratransit is to carry the rest that formal transit cannot cover may not necessarily be correct. From the perspective of urban transportation planning in practice, it may be more accurate to say that mass transit fulfills needs (volume, speed) that are no longer covered by paratransit. In addition, the understanding that paratransit is responsible for the mobility of people who cannot afford a private car or motorcycle should also be reconsidered. From an urban transportation planning perspective, a shift in people's use from public transportation, including paratransit, to private transportation is fundamentally undesirable. It has actually been observed that private transportation users choose to use flexible paratransit depending on the situation, and more attention should be paid to such advantages and potentials. Furthermore, even if paratransit is a captive mode for low-income groups, whether the fares are affordable for them is not the same story; since for-fire paratransit is labor-intensive, its fare level is highly dependent on the income level of the country or city. For the poor, in particular, it is an expensive mode of transportation; even in the case of collective paratransit, it is not inexpensive for minimum-wage workers, and in some cities, it is quite expensive. Moreover, the level of fares for longer distances to city centers is often not affordable for residents living on the outer edges of cities.

4.1.3.2 Flexible Service

The characteristics and benefits of being demand-responsive are described in Section 4.1.3.1. Informality, in the sense that there is less involvement by the public authorities, allows for flexible supply and operation in response to demand. In addition, even in areas with poor roads, transportation services are provided by small- and medium-sized vehicles. Paratransit may be almost the only means of transportation in informal settlements with large low-income populations. The use of small- to medium-sized, relatively inexpensive vehicles also creates low barriers for new entrants. Paratransit is also an adaptive transportation in emerging countries. Furthermore, it has the potential to be positively repositioned as a flexible and demand-responsive mode as a feeder to mass transit.

4.1.3.3 Entrepreneurial Business

Very small companies or sole proprietors operate many paratransit systems. They also create many jobs. They are basically established as businesses with little government involvement and no government subsidies. Countless small businesses or sole proprietors often form unions in an area or route. In some cases, the associations or cooperatives are self-regulating and controlled. Sometimes the cooperatives may set up and operate their own facilities for arrivals and departures. From the government's perspective, this provides mobility to people without subsidies. Many paratransit management groups and vehicle owners have an entrepreneurial spirit. While many aspects of paratransit remain old-fashioned, they may also be willing to introduce new technologies.

4.1.3.4 Immaturity of Services

Service uniformity and visibility are low, and there is generally no operational control or coordination. The different management actors in each vehicle result in competition in the market within an area or route, which further deteriorates the service. There is also a strong sense of territoriality among paratransit providers, and there are conflicts between driving crews and associations. Fares are often set by unions or regulated by the government in the case of route-based systems, but this is not always made clear to users, or drivers may demand higher fares depending on the situation. For-hire systems are often determined through negotiations with the drivers. In some cases, a fixed price is set by the association, or a meter is installed in the vehicle. However, it is also difficult to provide common fares or subscriptions for different modes. As for the means of payment, cashless payment is commonly introduced in some cases; for the for-hire type, such opacity is considerably eliminated in the case of shared mobility through apps provided by the platform. In some countries, harassment by drivers is also an issue.

4.1.3.5 Low Level of Technology and Infrastructure

A common situation in emerging countries is that it is difficult to invest in vehicles and services. The quality of vehicle maintenance tends to be low. Although this has been improving in recent years, pollutants emitted from vehicles with a low level of maintenance have a serious impact on air quality.

In addition, the aforementioned competition in the market causes aggressive driving in order to get as many customers as possible, and there are many road safety problems. In addition to those resulting from lack of funding, these problems stem from low awareness and skill levels among management and workers. Government regulations to improve these issues are often inadequate or ineffective. In addition, infrastructure for terminals and ranks for paratransit arrivals and departures is often not developed or underdeveloped by the government. Both route-fixed and for-hire paratransit vehicles often use on-street rather than off-street depots as their departure and arrival points. On-street departures and arrivals, as well as on-street boarding and alighting, tend to become bottlenecks for road traffic and cause congestion.

4.1.3.6 Low Working Environment

One of the major problems resulting from informality is the low working environment. Drivers hired by vehicle owners and managers are considered to have extremely low incomes, especially in the case of non-motorized paratransit that does not require skills. Vehicle owners who are drivers also tend to work long hours in order to earn more money (and to pay their vehicle loans). In some cases, there are no regulations regarding working hours, or if there are, they are not followed. Safety for workers as well as passengers is also an issue.

4.1.3.7 Value as a Cultural Icon

Apart from the issues discussed earlier, paratransit has developed uniquely under local conditions, and its existence and form may be recognized as part of the country's culture as a result. There are many examples of uniquely designed vehicles that can be considered icons of their cities. The Jeepney in the Philippines, a locally manufactured body replaced by a used truck chassis, is traditionally designed somewhat similar to a Jeep, with the owner adding his own decorations to attract customers. The jeepney has become one of the icons of the Philippines. While tricycle (three-wheeler) vehicles from Indian manufacturers are spreading globally, not only in South and Southeast Asia, but also in Africa, the Thai tuktuk vehicles are somewhat different in appearance. Outside of Thailand (including developed countries), Thai tuktuk vehicles have become icons that evoke the image of Thailand, whether used for tourism or placed symbolically in Thai restaurants. In addition, paratransit by bicycle, once common in Southeast Asia, has been transformed for tourists as a vehicle that offers a special experience. City tours in these vehicles, primarily for foreign tourists, are also popular. Although some of these vehicles have become less of a transportation mode than in the past due to the modernization of the vehicles and the transition of users to other modes of transportation, their soft power as icons of countries and cities may be rising rather than falling. Matatu in Nairobi has various graffiti painted on the body of the vehicle to attract customers by the owner, and inside the vehicle, not only graffiti but also speakers are installed to play various kinds of music. These are not meant to induce nostalgia but are considered trendy. There are many artists and craftsmen who paint graffiti on matatu, forming an industry. The graphic matatu, already part of the Nairobi cityscape, seems to have the potential to eventually be considered an urban icon.

4.2 VARIETY OF VEHICLES

As mentioned earlier, service types are broadly classified into route-fixed collective transportation (equivalent or close to bus service) and door-to-door individual transportation for-hire (equivalent or close to cab service). Vehicles are broadly classified into two categories: non-motorized and motorized. The non-motorized vehicles are bicycles with seats. In some parts of Africa, a passenger is carried on the back of a bicycle, or a driver pulls the vehicle (the earliest modern paratransit is a human-powered vehicle that utilizes bicycle spoke and tire technology (animal-powered vehicles are not covered here). In the motorized case, there are many varieties, including the following. Each country has its own name for them (Phun and Yai, 2016; International Association of Public Transport (UITP), 2024).

4.2.1 Motorcycle-Based

Motorcycle taxis are found in many countries. This is a passenger (or passengers) riding on the back of a regular motorcycle, with no special modifications. They have long been popular in Southeast Asian countries such as Vietnam, Thailand, and Indonesia, and have spread to other Southeast Asian countries and many African countries over the past few decades. Motorcycle taxis are rarely banned except in some countries where the motorcycles themselves are banned. Some countries have registration systems for vehicles and drivers. Since the 2010s, app-based shared mobility services have rapidly expanded in many countries. Other types of vehicles are three-wheeled motorcycles converted into three-wheeled vehicles with seating on the side or in the back. Some have a capacity of about six people (e.g., qingqi in Pakistan), which are used for route-fixed transit.

4.2.2 Three-Wheelers (Three-Wheelers, Tricycles, Autorickshaws, etc.)

This is a three-wheeled paratransit vehicle manufactured as a three-wheeled vehicle with a simple front wheel and steering wheel structure, making the vehicle inexpensive. Most of them have a capacity of 2–4 people. They have long been popular in South Asian countries including Thailand, Indonesia, and India. They used to be based on small two-stroke motorcycle engines, a source of air pollution. Today, most of them are CNG (Compressed Natural Gas) fueled, four-stroke gasoline-powered, or electrically powered. The small size of the vehicle makes it easy to introduce new technologies. Basically, they are for-hire individual transportation. There are also route-fixed services with a capacity of 4–8 people. The tuktuk of Thailand is known worldwide, but vehicles made in India are spreading worldwide. In particular, app-based shared mobility is rapidly expanding in many countries. In some countries, importers of Indian-made three-wheelers and app platformers collaborate to sell vehicles and recruit drivers. The service is spreading not only to Cambodia and other Southeast Asian countries, but also to a wide range of African cities.

4.2.3 Passenger Car-Based Vehicles (Capacity of 4–8 Passengers)

Passenger car-type vehicles are also popular among taxis. Basically, they are used without major vehicle modifications. Shared mobility, a service similar to taxis, is also mostly this type of vehicle.

4.2.4 Van/Wagon, Small Truck, or Pickup Truck-Based Vehicles (Capacity of 6–20 Passengers)

The original vehicle may be used as is, or the body and seating may be modified. Wagons made by Volkswagen, Toyota, and other brands (with a local capacity of 14–16 passengers) are very popular in lower- and middle-income countries in Africa and the Middle East as a route-fixed collective service using used cars because of their robust structures and ability to withstand severe use. Other emerging countries also use many wagons of this size. Smaller wagons (6–8 passengers) are also widely used because they can fit into narrow roads. Although few modifications other than the seats have been made, the doors may be removed for frequent boarding and alighting.

Some trucks can accommodate 6–12 passengers by modifying the bed of a small truck or pickup truck. These vehicles are often used because they are based on trucks and are powerful and sturdy.

4.2.5 Minibus (Capacity of 20–50 Passengers)

Many of the vehicles originally manufactured as minibuses are used almost as is. Others are modified by using a chassis from a used truck or other vehicle and placing a body manufactured in a traditional cottage industry factory (e.g., jeepneys in the Philippines). There are also new vehicles that combine the chassis of a medium-sized truck with the body of a local bus body manufacturer. Many minibuses in Asia and Latin America are of this type. In Kenya, ISUZU East Africa sells locally assembled new vehicles that use the bed of the truck as passenger seats, and these are widely used in matatu. Since the driver's cabin of the truck is left as it is, the driver's cabin and the passenger compartment are completely separated.

4.2.6 Medium and Large Buses (Capacity of 50 or More Passengers)

In addition to large bus vehicles made by new vehicle manufacturers, many are manufactured by body manufacturers who purchase chassis from global vehicle manufacturers and place them on bus bodies. In some cases, second-hand bus vehicles originally used in developed countries are utilized.

Thus, some vehicle types that are practically almost standardized worldwide are used as they are, some vehicles are modified locally, and some bodies are built new locally. Both types of vehicles may be new or used. The low local price, ease of maintenance, and specific local conditions determine the type of vehicle to be used. Since there are not many choices in vehicle types, they tend to be similar worldwide.

4.3 OPERATION AND REGULATIONS OF PARATRANSIT

4.3.1 Actors and Organization

Paratransit has a variety of actors, and their relationships vary from city to city. In many cases, they consist of a myriad of small enterprises or sole proprietorships, and other actors also perform their own limited roles in an attempt to maximize their profits.

4.3.1.1 Actors

The actors directly involved in paratransit business are listed in the following list. These actors are generally common to many paratransit systems around the world, but they are not always found in all paratransit systems, and they may also play more than one role (Baffi and Lannes, 2021).

- Driver
  The driver of a paratransit vehicle. They can be either owners or employees. In some cases, they only drive the vehicle, but in many cases, they are almost entirely responsible for the overall operation of the vehicle (maintenance, operation planning, etc.).
- Crew members (excluding drivers, but including assistants, conductors, and fare collectors)

They work with the driver to board the vehicle, and are responsible for touting, assisting the driver, and collecting fares. If the capacity of the vehicle is small, the driver alone performs these duties.

- Touts and dispatchers
  They provide information to passengers and touts at departure points (terminals and ranks), direct the departure of vehicles, and manage traffic. They may be assigned by the operator or the association, or they may be performed independently by private individuals who voluntarily perform these tasks and earn commissions from the association or drivers.
- Vehicle owner
  The owner of a vehicle. There are two major types of vehicle owners: those who are more like vehicle renters (whose objective is to make money from rental fees from drivers) and those who are more like individual operators. In the latter case, the driver may take on the role of operator to some extent, or the owner may be more involved in recruiting drivers, managing the vehicles, and operating the fleet. In some cases, the owner owns not one, but several or more vehicles.
- License holder
  The license here means a permit to operate a business, or the license to operate in a specific area or route. It may be "official" by the government or self-regulated by an operating association. There may be cases where the license holder is also the vehicle owner, where the operator or association holds the license, or where the license holder only holds the license and rents it out to vehicle owners or operators.
- Operator
  This is difficult to define. In paratransit, operations-related tasks are often divided among the drivers and owners, and in many cases, there is no clear "operator." In other words, what is referred to as an operator in paratransit is often more appropriately understood as a driver or owner in effect. Corporate organizations such as bus and taxi operators, which oversee operations as a single entity, are now semi-formal. Many of these have become corporate organizations through associations or a confederation of associations, which will be discussed next, or through mergers of these. There are also "paper companies" whose main purpose is to be officially registered, but which have no real operational control, when they are officially regulated by merging a myriad of sole proprietorships and small businesses and licensing only those organizations.
- Association
  An association is an organization of decision makers (drivers, owners, "operators") on the same route or in the same area. It is also called a union or cooperative in English. Originally formed voluntarily by

competitors, it establishes internal rules, sets order, and manages terminals and ranks. They also control entry and often charge a joining fee to those who wish to join. It is a window for external negotiation, and its important purpose is to preserve its own rights. Associations are the entities that perform this type of self-regulation. If there is no or very little governmental regulation, it tends to be a mafia-like organization as well. There can be conflicts and sometimes strife with other associations.

4.3.1.2 Relationships among Actors

Paratransit organizations can be understood as relationships among actors. First, the relationship between the actors directly involved in the operation should be understood, i.e., the relationship between the driver, owner, and license holder (Ferro and Allaire, 2015). When there is an explicit operator, these plus the crew and touts, all five roles are played by the operator (and it is a corporate organization with multiple or more vehicles). The infancy of paratransit's operating organization, business model, and services can be said to result from these five independent actors playing the roles of the operator, each acting under their own business interests. In the absence of a clear operator, the relationship between driver, owner, and license holder can be organized as follows. It is more common for the same paratransit in the same city to have a mixture of the following patterns.

- Separate driver, owner, and license holder (hired driver)
- Owner and license holder are the same, driver is separate (hired driver)
- The license holder is separate, and the driver and owner are identical (owner-driver)
- Driver and owner and license holder are the same (owner-driver)

The relationship between the hired driver and the owner also varies. Basically, the driver pays the owner a rental fee. In cases where the number of users and sales can be almost accurately tracked (ticket stubs, taximeters, cashless payment), the rental fee is linked to some extent to sales. In many cases, however, especially with cash payments, it is difficult for the owner to accurately track sales (even if the driver or fare collector pockets the revenue, the owner will not know). For this reason, a target system is used in which a fixed amount is paid to the owner from the sales, plus the driver pays for fuel, minor repairs, and police fines (or bribes), and the remainder is the driver's and crew's income. In the case of collective services, there is almost never a salaried system. This means that drivers try to make as much money as possible, which leads to poor service quality and low reliability. Dangerous driving and traffic disruptions due to long waits for customers on the road are common. Adjusting the number of trips and schedule-based operations are also difficult to implement because there is no benefit to the driver under the target system. The situation is much the same for owner-drivers. However, the fact that they do not have to be overly concerned about the targets and that they tend to treat their vehicles with care since they own them themselves may have a small effect on their driving behavior.

Associations are joined by vehicle owners or operators who pay a membership fee. In some cases, the association may operate a route or area by controlling the drivers, with operations-related tasks contracted out by the vehicle owner. In some cases, associations manage with more centralized control, while in others, they are simply councils where each owner has more power.

Associations are often in conflict with regulatory oriented governments. As a group representing the interests of the owners, they may refuse to change. On the other hand, the government, in the process of seeking to make paratransit more ordered and modernized, often encourages myriad "operators" to join or establish associations, or even to merge them. If decision-making and internal control can be established in some larger unit, it will be easier to coordinate with different modes of transportation and to improve and unify the quality of service. In addition, if the government wishes to make various approaches to paratransit, it can do so through associations. Associations are also important actors for the internal modernization of paratransit.

4.3.2 Operation and Cost Structure

4.3.2.1 Route-Fixed Transit

The operation of a route-fixed transit system depends not only on demand characteristics, but also largely on the relationships between paratransit actors and the cost structure. When operated by a myriad of private operators, it is very common to see "fill and go operations," especially on the origin side of the route (city center side, transportation hub side), in which vehicles depart when all seats are filled. It is an important role of touts and dispatchers at terminals to fill the seats as quickly as possible and have the vehicle depart. Although vehicles will depart even if they are not full, if they are going in the opposite direction of demand, the operation prefers to keep vehicles nearly full on the busiest sections of the line. In off-peak times, the association may establish a time interval for the service to generally depart regardless of whether it is full or not (this is the role of the dispatcher), but in many cases, the operator will stay longer until the seat is full, and the waiting time for users will be longer and more uncertain. One reason for this is that each individual operator is trying to maximize profits, so the mechanism is not designed to improve total service and increase revenues throughout the route and throughout the day. Another is the cost structure.

In developed countries, 60% to 70% of bus costs are driver costs, so it is rational to set the frequency of service (i.e., determine the number of drivers assigned) according to demand. In emerging countries, the proportion of fuel costs is relatively high. With the exception of some oil-producing countries, fuel prices do not differ significantly from country to country, but wage rates vary widely. As a result, the lower the income of a country, the higher the percentage of fuel costs and the lower the percentage of labor costs. Therefore, they prefer to operate with as full a capacity as possible to keep fuel costs per user low, even if it means longer waiting times for drivers and crews. Thus, in paratransit in low-income countries, terminals and ranks are full with crews of resting vehicles during off-peak hours, on the one hand. At the same time, vehicle costs are relatively high, so it is necessary to increase total revenues. This leads to long vehicle operating hours and the resulting longer labor hours for crews and poor maintenance (not taking the time to maintain the vehicles). On the other hand, during peak hours, the lack of transportation capacity results in long waiting times for users. Long queues at origin points and nearly full seats in the middle section of the route result in passengers not being able to board the vehicle. This is also due to the relatively large proportion of vehicle costs, which, unlike buses in developed countries, makes it unprofitable to ensure that vehicles are available to meet peak demand.

Additionally, corridors with high demand are also likely to have overcapacity, with many routes concentrated there. This is also a major cause of road congestion. This is because the myriad of individual operators, as well as associations and bus operators, prefer to set up routes that go directly to the city center in order to maximize the number of passengers using the route. This is a rational decision for profit-driven private operators, and it will require the involvement of public authorities to correct this. Paratransit is fundamentally demand-responsive, a characteristic that allows it to respond flexibly to people's mobility needs. However, as noted earlier, it is not necessarily demand-responsive when viewed microscopically at different times of the day and on different sections of the route.

The main reason why paratransit in emerging countries prefers smaller vehicles is because of the low per vehicle cost and small initial cost. In addition, even if the vehicle price per capacity is less costly for larger vehicles, the smaller capacity vehicles are more profitable because it is more difficult for larger vehicles to operate at full capacity and the off-peak service intervals would be longer (because they would have to wait until the vehicle is full). As a result, small- and medium-sized paratransit vehicles will operate extremely frequently in major urban corridors. There, unless the vehicles are full, there is virtually no waiting time. Competition often occurs in the same corridor between small paratransit and publicly operated buses with large vehicles. In this case, in the absence of some regulation of paratransit, the large bus routes would be less competitive, resulting in profitable paratransit and unprofitable large buses.

4.3.2.2 For-Hire Transit

For-hire transit is often conventional, with vehicles queuing up in spontaneous ranks at markets, large intersections, and other places where people congregate. Sometimes associations negotiate with the authorities to obtain "permission" (sometimes for a fee, sometimes for a bribe). This is the territory of each association. There are, of course, cases where vehicles flow empty on the streets, but in order to reduce fuel consumption, they often stop somewhere to wait for customers. In the case of app-based shared mobility, this does not require a specific rank, as the user calls it. As a result, the efficiency of collecting passengers is much higher than with conventional types, and in many cases, the driver's income is reportedly higher. The spread of share mobility in emerging countries has been more pronounced from traditional paratransit than from taxi-type passenger vehicles, plus the appearance of new paratransit in emerging countries. This is because the vehicles are inexpensive, allowing many drivers to get started, and the apps allow for high ride efficiency.

4.3.3 Regulations/Enforcement

The main regulations by public authorities may include the following:

- A: General regulations on vehicles, drivers (licensing), and traffic rules
- B: Various requirements for registration of operators, vehicles, drivers, and quality of service in the transportation industry
- C: Labor regulations for drivers and crews
- D: Registration and licensing of operating routes or areas and various regulations (e.g., entry regulations including number of vehicles and service-restricted areas)
- E: Fares
- F: Others: terminal and boarding point usage permits (on-road and off-road)

Not all of these regulations are imposed on all paratransit in a city. Some cities have almost no regulations at all. The content and intensity of regulations also vary. For example, a "vehicle," for example, may be registered as a commercial vehicle, have requirements for vehicles in use and maintenance, and may have additional regulations on the number of vehicles and their entry. These can be divided into those that apply broadly to ordinary vehicles and those that apply to those operating in the transportation industry. Furthermore, depending on the strength of the regulation, there are various types of licenses/permits (which are determined by meeting certain requirements, plus the existence of prior applicants or competitors), approvals (which are basically accepted if certain requirements are met), registrations (which are accepted upon submission and completion of their records), notifications (only submitted), and so on. Informality from a regulatory perspective is a situation where some of the aforementioned regulations do not exist, or are weakly regulated, or even if regulated, are not effective. From the perspective of improving the quality of transportation services, ensuring transparency, protecting workers, and ensuring safety, it is desirable that these rules be established and followed. The operators' awareness of compliance and the capacity of public authorities should be improved.

In many cases, self-regulation, mainly by associations, is often the way to go. Particularly, in the case of regulation D (regarding entry) and regulation E (regarding fares), where there is no or loose administrative regulation, some kind of self-regulation will apply. In addition, associations may set requirements regarding the quality of service (e.g., vehicle requirements, provision of information to users).

In general, in route-fixed paratransit, the administration often requires some kind of permit, registration, or notification for D (entry); often approves E (fares; often a political issue for low-income groups), and F is approved by the administration's road department or the police. Some permits are formal, while others are tacitly approved by the local officials (by paying a bribe rather than a fee). In order to improve the quality (including safety aspects) and modernization of paratransit transportation, the regulation of B and C and their effectiveness are crucial first. In other words, modernization of operations and formalization of employment by operators is first and foremost important. The general strategy in doing so is to merge the myriad of individual operators and encourage them to belong to associations, and then further merge associations and cooperatives. This will encourage the endogenous modernization of the operation itself, while at the same time reducing the number of operators subject to licensing and increasing the effectiveness of government regulation and management. In addition, if integration as an urban transit system is the goal, the reform of regulations regarding the entry of D and the incentives provided to individual operators through the expenditure of some form of operating subsidy will be an issue.

For-hire paratransit is often less regulated or almost non-existent than route-fixed paratransit. However, regulations on vehicles, drivers, and quality of service, as well as labor regulations, are important to improve the quality (including safety) and modernization of transportation. As the number of app-based shared mobility and its rapid conversion from existing paratransit is increasing, many argue that while quality is guaranteed by the platform, it is not sufficient in terms of worker protection. The business model in which the driver bears all risks for the operation has been pointed out as a problem even in developed countries. This is a dynamic change that needs to be carefully monitored.

4.4 COORDINATION AND INTEGRATION OF PARATRANSIT WITH INSTITUTIONAL MASS TRANSIT SYSTEMS

4.4.1 What Formalization and Integration Mean?

It is important to modernize paratransit and, at the same time, to integrate paratransit with so-called formal mass transit systems such as BRT and MRT (mass rapid transit) as an urban public transportation system. And there is no doubt that this is the global direction (Ferro and Allaire, 2015; Jennings and Behrens, 2017; Cassius et al., 2021; United Nations Development Programme (UNDP), 2024). The reminder here is that there is no single definition of what formalization and integration mean. This is because there is no single desirable model of urban public transportation system, depending on how urban public transportation has developed so far and on the local conditions of a given city or country.

Before that, rethink informality. In the past, paratransit was regarded as informal transportation, which was either illegal or necessary but undesirable. It was then believed that the direction of urban public transportation development was to eliminate its informal nature and establish "formal" transportation. However, this has come to be seen as not necessarily a uniquely universal measure for two reasons: first, the means of eliminating the paratransit sector, which creates many jobs, would leave its vulnerable employees and users in an even more difficult position. The second, and this is more substantial from a transportation planning perspective, is that "formalization" or "integration" could also result in the loss of some of the benefits of paratransit. For example, the advantages of service being commercially viable without subsidies, service being improved by the operator's profitability mindset, and more frequent service being provided by smaller vehicles may be lost, resulting in less convenience for the greater cost.

Formalization of paratransit means modernization of paratransit operations and professionalization of those involved, with the goal of improving the quality and transparency of transportation services, protecting workers, and ensuring safety, and as a result, shift from private transportation to public transportation. In order to achieve these goals, the formalized operational form (in the case of fixed-routes) and the type of system that is most desirable (licensing, concession, franchise, or contract (with subsidies)) will depend strongly on the current state of the city's paratransit, the network, demand structure, and so on.

As for the integration of paratransit with other public transportation modes, the ultimate goal is not necessarily to integrate fares, operating schedules, and operating entities as in the European Transit alliance (i.e., almost synonymous with the introduction of a contract system for paratransit).

The process of introducing BRT requires coordination between BRT and existing paratransit, and in many cities that have actually introduced BRT, BRT and existing paratransit have integrated fares and operations in a great many cases. However, first of all, the joining of an association of individual paratransit operators or the merging of associations and operators is one process of integration, and in some cases it may be the goal. Even if mass transit and paratransit, including BRT, are not fully "integrated," they can often be considered to be sufficiently integrated as an urban public transportation system when they are physically connected and "coordinated" to form a trunk line and branch line network.

4.4.2 Modernization Models of Paratransit

Modernization and reform of paratransit can be broadly classified into the following two models:

4.4.2.1 Centralized and Institutional Integration Model

This is one in which a single organization for public transit integrates public transit on a contract basis, ranging from one in which the BRT operating organization integrates its feeders to one in which almost all public transit in a city, including BRT, is integrated. The integration of paratransit as a BRT feeder on a contract basis is more common in Latin American cities (Tun et al., 2020) and has recently been done in Jakarta, for example. Bogotá has been building and integrating feeder buses since the BRT began its service in 2000, gradually converting all informal general buses in the city to semi-formal buses beginning in 2012, and integrating all buses on a contract basis in 2021. The integrated buses were equipped with low-emission bus vehicles and EVs, and the full formalization had significant social benefits (Rodriguez et al., 2016). However, the integrated general buses, which were formerly operated by private operators on a self-financing basis, are heavily subsidized currently, while the satisfaction level of the users remains unchanged due to the need to make transfers additionally (Rodriguez-Valencia et al., 2023). In Jakarta, a portion of angkot, the original route-fixed conventional paratransit, has been converted to contract-based minibuses (mikrotrans) as feeders for the BRT (TransJakarta) and integrated with TransJakarta. Users will be able to use IC cards operated by Jaglngko for free of fare. This has increased the area covered by the TransJakarta system from 42% to 81.9% (Institute for Transportation and Development Policy (ITDP), 2021). Although this method seems "ideal," it requires a large amount of subsidy and there are generally significant difficulties in coordinating with existing paratransit operators and owners. Especially when existing paratransit associations, cooperatives, and operators are weak, coordination is nearly impossible. When implementing this measure, there are various considerations, such as whether all or some of the integration should be done, whether contract-based integration or some other method is preferable, and what type of vehicle is preferable when replacing paratransit.

4.4.2.2 Decentralized and Coordination Model

This is a professionalization and modernization of the paratransit sector. It makes good use of the characteristics of paratransit, especially its self-financing nature, while reducing the negative externalities of paratransit and promoting coordination with other modes of public transportation. In particular, it sets requirements (facilities, safety) for operators (owners) to comply with, and raises the level of technology and awareness of compliance. For example, in the Philippines, the Public Utility Vehicle Modernization Program is underway to replace paratransit vehicles, including jeepneys, which are mostly used vehicles that have been converted, with vehicles that have higher environmental and safety performance. Paratransit jeepneys, many of which are individual operators, are required to join an association as a condition for licensing. Government subsidies for vehicle replacement are available but very limited, while the price of vehicles that meet this program is extremely higher than conventional jeepneys, so the refusal of some jeepney owners is reportedly very strong. All matatu in Kenya are required to belong to an association (SACCO) and are subject to various notifications. Although a system of registration, notification, and licensing exists, the system of regulation of the quality of service and safety and its inspection is poor, and many SACCOs, owners, and crew members lack professionalism.

This method is superior when there is no mass transit system, such as BRT, or when the scale is small, and when the paratransit sector is profitable and has a very large number of operators. This type of scheme is sometimes referred to as a hybrid scheme of formal mass transit and informal (or semi-formal) paratransit. For the purposes of this chapter, it is a hybrid of a contract-based trunk service and a profit-based feeder or non-trunk service. However, some form of financial subsidy or incentive is essential if fleet replacement is to proceed. Public intervention in the development of infrastructure, such as terminal construction, is also essential.

4.5 FUTURE PERSPECTIVES ON PARATRANSIT

Paratransit in emerging countries was once perceived as informal, sometimes illegal, and low-quality transportation. In reality, however, many of them operate under some form of regulation or guidance from public agencies, and have the advantages of flexibility and commercial operation without subsidies, and have the potential to be urban public transportation systems and actually have many users. Taking advantage of these strengths, in order to reduce the negative externalities of the current situation and also to promote a shift from private transportation, it is important to improve the quality of transportation services, ensure transparency, protect workers, and ensure safety. The way to do this, especially in route-fixed paratransit, is to move from maximizing the interest of individual operators to encouraging them to act as corporative operators and acquire the capacity as a professional group to optimize their own transportation on the route or area as a whole. This is not necessarily centralized integration and institutional formalization led by public authorities of the developed world type. It is a reform that takes advantage of the strengths of paratransit, even if it is a modest one: the integration of services and technological modernization through coordination by modernized and decentralized actors. Establishing such a model is required.

Finally, mainly for-hire paratransit will be discussed. The transformation of conventional paratransit into app-based shared mobility has improved the quality and transparency of service and increased driver income in emerging countries (United Nations Development Programme (UNDP), 2024). Although there are conflicts with existing paratransit and cab services, quality assurance by global or local platforms and linkages with other services are being offered. For shared mobility using private vehicles, a number of emerging countries have introduced different regulations (as hire cars), which are different from those for cabs. In terms of quality assurance, for example, Grab in Vietnam provides safety training, uniforms, helmets, and insurance coverage and payment for aspiring motorcycle taxi drivers. The share of the driver's income is estimated to be about 50%, which means that the platform is spending money to improve the quality of the service. At the same time, shared mobility is a service provided by a sole proprietor, and from this perspective, it is truly informal. The business model of providing a match between individual drivers and users is innovative because of its informal nature. However, especially in developed countries, the working environment of gig workers is problematic, and this structure is basically the same in emerging countries. It has not become a major issue in emerging countries, simply because it is relatively less problematic than other occupations. It can be said that the integration and formalization of services by platformers and informalization as labor is increasing, preserving informality as a means of transportation. Furthermore, there is still debate over whether the spread of shared mobility will lead to a reduced ownership and use of private transportation by people, and thereby reduce total traffic volume, or whether it will accelerate an increase in traffic volume. More multifaceted evaluation and monitoring of shared mobility in emerging countries is required from the perspective of urban transportation planning.

REFERENCES

Baffi, S., & Lannes, J. P. (2021). Understanding paratransit: Defining and diagnosing paratransit for sustainable mobility planning. *MobiliseYourCity*. Retrieved from https://www.mobiliseyourcity.net/sites/default/files/2024-05/Paratransit_Tool%20I_ENG_2704_compressed.pdf

Behrens, R., Chalermpong, S., & Oviedo, D. (2023). Informal paratransit in the Global South. In R. Behrens, S. Chalermpong, & D. Oviedo (Eds.), *The Routledge handbook of public transport*. Routledge.

Cassius, S., Deeb, N. E., Sorour, M., & Turner, S. (2021). Future of paratransit and shared mobility: Mapping report. *Institute for Transportation and Development Policy (ITDP)*. Retrieved from https://itdp.org/publication/future-of-paratransit-and-shared-mobility-mapping-report/

Cervero, R., & Golub, A. (2007). Informal transport: A global perspective. *Transport Policy*, *14*(6), 445–457. doi:10.1016/j.tranpol.2007.04.011

Ferro, P. S., & Allaire, J. (2015). *Paratransit: A key element in a dual system*. CODATU. https://www.codatu.org/wp-content/uploads/2024/03/transports-collec-artisanal-v03ecran-en.pdf

Institute for Transportation and Development Policy (ITDP). (2021). *Lessons learned from Jakarta's journey to integrated and resilient transport systems*. https://itdp.org/publication/lessons-learned-from-jakartas-journey-to-integrated-and-resilient-transport-systems/

International Association of Public Transport (UITP). (2024). *Description, roles and functions of paratransit (Policy Brief)*. Retrieved from https://www.uitp.org/publications/description-roles-and-functions-of-paratransit/

Jennings, G., & Behrens, R. (2017). The case for investing in paratransit: Strategies for regulation and reform. *Volvo research and educational foundations (VREF)*. Retrieved from https://www.mobiliseyourcity.net/sites/default/files/2021-07/Investing%20in%20Paratransit%20-%20Jennings%26Behrens-August2017.pdf

Phun, V. K., & Yai, T. (2016). State of the art of paratransit literatures in Asian developing countries. *Asian Transport Studies*, *4*(1), 57–77. doi:10.11175/eastsats.4.57

Rekhviashvili, L., Kębłowski, W., Sopranzetti, C., & Schwanen, T. (2022). Informalities in urban transport: Mobilities at the heart of contestations over (in)formalisation processes. *Geoforum*, *136*, 225–231. doi:10.1016/j.geoforum.2022.07.012

Rodriguez, C., Peralta-Quiros, T., Guzman, L. A., & Cárdenas Reyes, S. A. (2016). *Bogota's bus reform process: Accessibility & affordability effects, lessons learnt & alternatives to tackle informal services*. World Bank. https://thedocs.worldbank.org/en/doc/697341488490154585-0190022017/

Rodriguez-Valencia, A., Rosas-Satizaba, D., & Hidalgo, D. (2023). Big effort, little gain for users: Lessons from the public transport system reform in Bogotá. *Public Transport*, *15*(3), 411–433. doi:10.1007/s12469-023-00300-5

Tun, T. H., Welle, B., Hidalgo, D., Albuquerque, C., Castellanos, S., Sclar, R., & Escalante, D. (2020). Informal and semi-formal services in Latin America: An overview of public transportation reforms. *Inter-American Development Bank (IDB)*. Retrieved from https://publications.iadb.org/en/informal-and-semiformal-services-latin-america-overview-public-transportation-reforms

United Nations Development Programme (UNDP). (2024). *A closer look at informal (popular) transportation: An emerging portrait*. Retrieved from https://www.undp.org/acceleratorlabs/publications/closer-look-informal-popular-transportation-emerging-portrait

Chapter 5

Bus and BRT

5.1 HOW BUSES ARE USED IN EMERGING COUNTRIES?

Fumihiko Nakamura

5.1.1 Definition of Buses

This document deals with buses that are intended for daily use within and beyond a city, rather than buses that transport passengers over long distances between cities. Definitions of bus vehicles vary slightly from country to country. For example, in Japan, a vehicle with a capacity of 11 or more passengers is called a bus, regardless of how it is operated, while a vehicle with a capacity of 10 or less passengers is not called a bus but a cab, even if it provides service with routes and stops (technically, a "shared-ride cab" means shared transportation by a cab vehicle). Thus, the legal definitions differ slightly from country to country (Cervero, 1997). In this document, we will refer to the entire shared-ride transportation as a bus, regardless of the size of the vehicle.

On the one hand, not only in emerging countries but also in developed countries, urban space expands, so it is no longer possible to reach the cities in time on foot. On the other hand, as the cities become more densely populated to a certain extent, especially owing to the bulk demand for commuting to work and school during certain hours, systems to transport a certain amount of goods in bulk become more widespread. Although the history of urban development and the speed of development differ, it can be said that the process is the same in both emerging and developed countries. One service that can still be seen in some cities in emerging countries is a service in which an agricultural tractor is connected to a cart, and the residents are carried on the cart. This could be said to be the prototype.

Services that operate on converted truck beds still exist in many emerging country cities. They, too, should be classified as buses regardless of the form and dimensions of the vehicle. Needless to say, in emerging countries, there are not only such converted trucks, but also so-called buses with a capacity of several dozen or more people, which are imported used or new vehicles

 DOI: 10.1201/9781003467403-6

from developed countries. While used chassis and engines are imported, the bodies are manufactured in their own countries.

5.1.2 Management of Urban Buses

There are three main types of urban bus operations. The first is a very spontaneous local service that has been practiced, so to speak, as a way of life. Many of them still remain in many cities and are still used in close contact with the community even today with the increase in the number of automobiles. Many of them are modified vehicles, and thus have emission problems, risk of breakdowns, and so on. Many of those that have persisted over the years are organized to some extent and in this sense are well managed.

The vehicles used are generally small, with a capacity of no more than 30 passengers at most. Although the number of cars and fares are controlled, in almost all cases, the fare income from the passengers carried that day goes directly to the crew's salary, the so-called commission system. For this reason, a large number of vehicles come to railroad stations, commercial areas, or areas near markets, depending on the time of day, and the crews fight to attract passengers. While this is interesting from a tourist's point of view, it also poses many problems, such as congestion on surrounding roads, risk of accidents due to excessive competition, and the fact that there is no timetable and there is no way to know when a train will depart, leading to low punctuality and reliability.

The second is a form of bus service in which the entire region has a large number of routes and is managed by a management organization such as a transportation bureau attached to the municipality. In some cities, the entire system was imported from the colonial sovereign state, while in others, it was set up independently. Not many cities have this type of transportation authority. In Thailand, the Bangkok Metropolitan Administration is entirely managed by the Bangkok Metropolitan Administration of Transport. Some of the Bangkok Metropolitan Administration's routes allow small bus operators to operate on the same routes, as they are not able to provide transportation using only their own fleet of vehicles. Many of these operators have one or a few small vehicles. In some suburban areas, truck-modified vehicles operate as buses under the approval of the Bangkok Metropolitan Administration of Transport. These vehicles are assigned a system number, just like the buses operated by the Bangkok Metropolitan Administration, and the number is displayed on the vehicle. Although this is a system number, it can be said to have the same characteristics as a business management number.

In the Ekachai district of western Bangkok, there are about 1,500 small buses, called "Si Lo Lek," which are converted from light trucks, operating at high frequency. These buses are not recognized as buses by the Bangkok Metropolitan Administration and do not have a system number. The "TAXI" sign is displayed on the upper front of the bus. Although this service is not recognized as a bus service, it is a convenient service that is closely connected to the local area and is used by many people. Not only does it provide access to the train station connected to Ekachai Road and commercial facilities located along the street, but it also allows passengers to get on and off anywhere on Ekachai Road and runs nonstop once it reaches capacity, making it a convenient service that is never too slow and has taken root in the community. The Si Lo Lek on Ekachai Street falls into the first category. However, it is a very special case because it is managed very well by the local leaders, so there is almost no risk of congestion or accidents as mentioned earlier.

The third is a form in which a number of large private operators cover the entire area, backed by the financial power of commercial capital such as overseas Chinese. Hanoi and Ho Chi Minh City in Vietnam and Yangon in Myanmar are operated by several large bus operators. Because of some kind of coordination, there is no lawless competition, but they are separate operators.

5.1.3 Use of Urban Buses

On the one hand, in cities without subways, elevated railroads, monorails, new transit systems, and so on, buses are the main form of public transportation, regardless of the size of the vehicles. In many cases, its fares are controlled to low values, partly for political reasons, that is, politicians' intention to gain popularity among low-income people, as it is an important vehicle for low-income people. In some countries, the poor who cannot even ride its buses walk long distances. On the other hand, in Bangkok, where an elevated track system is being developed, its fares are sometimes 10 times higher than those of buses, and low-income groups continue to use buses. As a result, bus routes parallel to the track system remain firmly in place and there is a need for them.

In any case, buses play an important role in large cities in emerging countries, where there is a wide range of income levels, the size of cities is expanding, railways are not necessarily well developed, and neither driving licenses nor automobiles are widely available.

5.2 INTRODUCTION OF BRT

Fumihiko Nakamura

5.2.1 Pre-era of BRT

The term BRT began to be used around the world around 1996. Like other urban transit terms of this type, it has spread from the United States BRT stands for Bus Rapid Transit, an American term that refers to a rapid transit service that utilizes buses, which are fast-moving (fast here meaning faster than automobiles in congested road spaces), fixed-route transit vehicles. The term "bus" is used in the United States.

Of course, even before the term "BRT" was coined, there were already examples of buses being used in cities all over the world: In terms of mass transit, a bus convoy system in Porto Alegre, Brazil (a system that transports 360 buses per hour, or nearly 30,000 people per hour, by operating a convoy of six buses at one-minute intervals and using a long platform that allows six buses to arrive and depart simultaneously), a system that uses a three-body bi-articulate bus system (a bus convoy system in Porto Alegre, Brazil), and a system that transports 15,000 passengers per hour (a trunk express bus system in Curitiba, Brazil) by using bi-articulate buses connected to three car bodies and by devising boarding and disembarking points (Yabe et al., 2006).

In terms of speed and punctuality, Istanbul and Taipei, among others, had already introduced bus-only lanes in the middle of roads. Examples in developed countries include bus-only tunnels in Seattle and New York, as well as Runcorn in the United Kingdom and Almere in the Netherlands, which introduced bus-only roads when developing new towns (Nakamura et al., 2016).

5.2.2 Image of BRT

Nevertheless, the term BRT has attracted attention since 1996 for several reasons. First, the sound sequence can be contrasted with LRT (Light Rail Transit: a next-generation tram-dimension track system). With streetcars under review in many cities, especially in developed countries, this was in line with the intention of appealing as an alternative that is cheaper than LRT, has a shorter construction period, and is flexible enough to cope with various uncertainties. International aid agencies for emerging countries, led by the World Bank, also paid attention to BRT. In addition to the fact that BRT can introduce a mass transit system with superior speed and punctuality at a lower cost than track-based systems, it can also dispel the image of conventional buses in almost all emerging countries, which have a strong negative image of cleanliness, road safety, and on-board security, as well as other negative aspects such as pollution and bad smells. Therefore, the name "BRT" was chosen in order to wipe out the image of conventional buses and to emphasize that it is a new transportation system that is different from conventional ones. The bi-articulate buses in Curitiba, Brazil, described in the following section, were historically referred to as "surface metro" before the term BRT became popular. The claim was that, although mechanically the vehicle was a bus, the system was not a bus at all, but rather a subway running on the ground.

The BRT system that brought this image of BRT to the world's attention was the TransMilenio system in Bogotá, the capital of the Republic of

Colombia, which began operating in 1999. The details of this system are described in the following sections, but it is an example of a system that learned a lot from Curitiba and that was advanced in terms of mass transportation and rapid delivery, giving not only the citizens of the city but also the world a thoroughly different image from conventional buses. After the Bogotá case, a system called BRT was introduced all over the world, following in its footsteps. Although some countries, such as Japan, have misunderstood the greatness of this system, many emerging countries have started to consider the possibility of a system similar to Bogotá's. In China, the Bogotá BRT system started from Kunming in 2007 (the city's central travel lane was introduced in 1999), and in no time at all, BRT was introduced in many cities in the country.

ITDP, an international non-profit organization that promotes urban transportation policy, has created a database of BRT case studies from around the world, has published the guidelines, and has started BRT award activities. ITDP's activities (ITDP, 2007) are undoubtedly contributing to BRT considerations in emerging countries in a variety of ways.

5.3 BRT MANAGEMENT IN BOGOTÁ, COLOMBIA

Fumihiko Nakamura and Yurie Toyama

5.3.1 Why Should We Learn the Experiences in Bogotá

BRT often attracts attention for its articulated bus vehicles and center-running bus-only roads. In fact, the sight of large, clean bus vehicles designed by European manufacturers traveling at high frequency and high speed down the center of a busy road are enough to dispel the image of conventional buses. However, having a BRT is not as simple as just dedicating the center lane of the road to buses and buying new vehicles. In many cities, there is a considerable amount of existing bus service, and there are many issues to be overcome, such as how to reconfigure this service and how to handle the uncontrolled development and expansion of the city and the corresponding growth in demand for motorcycles and cars. The case of Bogotá is presented here because it is an example of a city that has learned from Curitiba and greatly evolved from it, and is therefore highly worth learning from.

5.3.2 Bogotá Learned a Lot from Curitiba, Brazil

Before we get into the case of the City of Bogotá, we will briefly touch on the case of Curitiba. Bogotá City has learned a lot from Curitiba City, including some antithetical elements.

As mentioned earlier, in Curitiba, based on the 1966 Master Plan, five structural axes were established within the city, and roads were developed using the reduction in the amount of land converted through the division of residential land, and bus routes were established. The 10 operators in the city were then separated, and their operations were taken over by the City Corporation. As a result, all bus routes in the city were, so to speak, defined in the city plan. In fact, all residential construction is prohibited in areas that are more than 500 meters from bus stops. All fare revenues are paid to the public corporation, which then distributes the money to each operator in proportion to the total distance traveled. This means that bus operators no longer aim to attract a large number of passengers, as in the past, but to run their buses safely. The amount of passengers on a bus route is no longer the responsibility of the operator, but rather the responsibility of the city, or in other words, the city planner.

Curitiba has grown as an industrial city centered on the Volkswagen automobile plant, and has become the fifth most populous city in the entire Brazilian federation. As the number of registered vehicles increased with economic growth, road congestion worsened. At the same time, the number of low-income residents, including those who had migrated from neighboring cities, increased, and mass transportation by bus became an issue. Under these circumstances, the stilt bus stops and bi-articulated buses were developed with the support of Volvo, a Swedish bus vehicle manufacturer. The high-floor bus stops are designed to shorten the time required for buses to stop at a bus stop. The platform is the same height as the floor of the bus interior, and the bus vehicle is equipped with a crossing board, which allows passengers to get on and off the bus without any steps or gaps. In addition, the bus conductor was eliminated and an attendant was stationed at the bus stop to collect fares in advance, thereby reducing the time and effort required to pay fares when boarding the bus and eliminating delays caused by such delays. This system was first introduced in 1991 on non-stop buses (buses that increased speed by making very few stops along the way). The high-floor bus stop became known as a tube bus stop because of its cylindrical design shape. The next year, 1992, saw the introduction of bi-articulate buses with a total length of 25 meters and a capacity of 280 passengers, according to calculations based on Brazilian federal law at the time. With five doors on each side, this bus had a short boarding time and high frequency of operation thanks to the introduction of tube bus stops. According to initial calculations, everyone could get off the bus in 23 seconds, which means that even if 280 people get off and 280 people get on, it would take less than one minute to get on and off.

In addition to this, the buses in Curitiba had put off improving their information and communication systems. Since there were attendants at tube bus stops, bus-only roads were fully equipped, and loop coils were used to detect buses at major intersections and control signals to extend the green time in the direction of bus-only roads, it is assumed that there was little interest in real-time operation management and other such systems.

5.3.3 The Need and History of BRT in Bogotá

Bogotá, the capital of the Republic of Colombia, is located at an altitude of 2,600 m above sea level. It is a dense metropolis with an urban population of 8.2 million (2019 Census) and a municipal area of 1,587 km^2.

Meeting the mobility needs of this large city requires a mass transit system: before the BRT, Bogotá's public transportation system was almost exclusively a minibus service provided by privately owned bus companies. These minibuses faced challenges such as insufficient number of vehicles, inefficient route networks, lack of administrative oversight, poor working conditions for drivers, and lack of consideration for users.

To address these issues, measures such as Car Free Day and Bicycle Paradise (Ciclovia) were implemented to encourage people to walk and cycle as well.

In 1998, the petrol tax was increased and restrictions on the flow of cars into the city center through number plates were implemented. In 1999, the BRT system, TransMilenio, began operating.

Bogotá's BRT was recommended in Japan International Cooperation Agency (JICA)'s urban transportation studies in the 1990s, alongside studies on the construction of a metro and a motorway, which led to the original proposal for construction.

Bogotá, like other large South American cities, is based on a road system on a grid. The main roads radiate out from the city center, and the dedicated roads on which the BRT trunk lines run are located around the trunk sections. The linking of trunk lines and branch lines at the terminals is learned from Curitiba, Brazil, but what differs from Curitiba is the thorough focus on the role and design of mass transit.

5.3.4 BRT Specifications in Bogotá

Bogotá's BRT system, TransMilenio, has 138 stations (2023) on 11 corridors with 99 trunk lines (2023) and 106 branch lines (2023) for a total system length of 114.4 km (2023). Peak passenger volume is 49,000 passengers/hour/direction (2018), transporting 1,840,000 passengers per day (2023).

5.3.5 BRT Innovations in Bogotá

5.3.5.1 Infrastructure and Urban Space Associated with BRT

When arriving in Bogotá by plane and heading out into the city, the TransMilenio is always visible, and the scale and urban presence of the dedicated TransMilenio road and station are astonishing. In order to realize

Figure 5.1 TransMilenio's typical station with overtaking lane

Source: (Photo taken by Yurie Toyama)

high-speed, high-frequency operation, each station on the main line has an overtaking lane (Figure 5.1) to enable departures at any time (i.e., without waiting for a bus to depart from the station ahead) and express operation, and the platforms are covered with glass and roofs, just like railway stations. The stations are equipped with security cameras and security guards to ensure a safe and secure environment for waiting buses. The creation of secure public spaces throughout the city has had a significant impact on improving the safety of the city as a whole, as Bogotá was known worldwide not so long ago for its insecurity.

It is not hard to imagine that the construction of these richer infrastructures would have cost a considerable amount of money. In particular, dedicated running space for mainline buses has been created by widening the width of conventional arterial roads to create space. They are being built there at Colombian state expense and with petrol tax revenues from the city of Bogotá.

This presence of TransMilenio infrastructure in the city has had various secondary effects. For example, the driving space in the city center could be converted into a substantial transit mall, with areas where ordinary vehicles could not flow in, allowing buses, pedestrians, and cyclists to coexist (Figure 5.2). In addition, the station serves not only as a transportation node, but also as a venue for Cultural Friday, a cultural and artistic event that makes use of the space in front of the station on Fridays. Orchestra concerts, children's events, and other events are used to be organized. Other events were held in the station, such as a mini-library, painting exhibitions, and photography contests (abandoned as of 2025).

Figure 5.2 TransMilenio in City Centre

Source: (Photo taken by Yurie Toyama)

5.3.5.2 BRT Operation and Management

The planning, operation, and management of the TransMilenio BRT system in Bogotá is carried out by the TransMilenio Corporation, which is 70% funded by the City of Bogotá and 30% by each of the central government ministries. The construction and management of the infrastructural facilities, such as the running track, stations, pedestrian bridges, and parallel cycle paths, are also carried out by the TransMilenio Corporation.

Under contract to the TransMilenio Corporation, seven private operators operate trunk bus services, six private operators operate branch line bus services, and two other private operators collect fares at stations.

Fare revenues are once centralized in the Corporation, with the trunk bus operators paid an amount proportional to the distance travelled, the branch line bus operators paid an amount proportional to the number of passengers carried, and the fare collection operators paid an amount proportional to the number of tickets sold.

Bogotá's branch-line buses mainly run in low-income neighborhoods, and almost 90% of their users are low-income people. Therefore, integrated with welfare policies, branch-line buses are fare-free. According to surveys, more than half of branch-line bus users transfer to mainline buses at the terminal, so it can be said that the income from mainline buses internally subsidies the branch-line buses.

All buses are equipped with a bus locator system, which provides location information and, at stations, there are automatic ticket gates to monitor the number of passengers. Maintaining the speed, safety, and punctuality of trunk line buses is a particularly important mission in order to prevent citizens from leaving the bus system. For this reason, the evaluation item for operators of trunk bus services does not include the number of passengers transported, but instead they are paid solely on the basis of pre-contracted transportation distances (which can be calculated using the system and timetable), with incentives paid according to performance such as the number of accidents and delays. This system maintains the relationship between government and the operators, which guarantees high service quality.

5.3.6 Learning from Bogotá

In introducing TransMilenio, the City of Bogotá learned from Curitiba the separation of operation and management, the use of high-floor bus stations (which are so large that they are more like stations than stops, and will henceforth be referred to as bus stations in this chapter), and the use of articulated buses. Moreover, the need to provide a considerable amount of transportation made it possible to operate at 10-second intervals by making the bus station the center of the system with two lanes on each side and introducing express operation that passed through several stations along the way. In Bogotá, unlike Curitiba at the time, bus priority in signal control at intersections was not realized (even in Curitiba, bus priority control at intersections was removed after 2004, following a policy shift from a policy emphasizing buses). In addition, since Curitiba has long been a city with poor public safety, a control center has been established for real-time bus operation management and security management at all bus stations, with TV cameras monitoring and guards stationed at stations.

There are three additional ways in which Bogotá's TransMilenio differs from Curitiba's. The first is the transfer process from the existing bus system. While Curitiba placed all bus operators in the entire city under the umbrella of the city corporation, Bogotá incorporated the operators of the areas parallel to the TransMilenio routes and branch lines, as opposed to the myriad of privately owned bus operators. Therefore, the old minibuses remain in some parts of the city as well as the TransMilenio. It is said that becoming a TransMilenio crew member was a very popular career change from the existing operators because of the advantages such as a maximum of 6 hours of driving time, guaranteed salary, and training in Paris and other cities. This focus on crew quality assurance and management is a major differentiator between TransMilenio and conventional buses.

The next distinctive feature of the project is the securing of financial resources for the widening of the road for the installation of a dedicated bus route. In the

Bogotá metropolitan area, the gasoline tax is set at a high level, and the revenue from this tax is used for the construction of infrastructure for the TransMilenio, which is an urban transportation improvement project. This has allowed the construction of more than 80 km of two-lane, single-lane, bus-only roads.

Another point is the linkage with urban planning. The city of Bogotá does not have a structural axis of development. To begin with, the city has a system that limits the areas where people can live according to their declared income, and low-income residents are relegated to the suburbs. TransMilenio directly connects these suburban centers to the city center. Bus terminals in the suburbs are connected to urban activities, such as commercial facilities or district offices, have large plazas, and hold events on weekends, but they are not connected in a TOD-like manner, as is the case in Curitiba.

As mentioned earlier, the TransMilenio in Bogotá City has achieved a transportation capacity comparable to that of a railroad and a high level of security that even the white-collar class uses, with the aim of thorough mass transportation and image elimination. Coordination with existing operators, assurance of crew quality, thorough use of information and communication technology, and a four-lane bus-only road made possible with gasoline tax funds are all examples of the unparalleled power of this system. TransJakarta in the Jakarta metropolitan area in Indonesia is highly regarded for its construction of 170 km of bus-only roads in a short period of time with the support of ITDP, but it falls short in terms of how much it has learned from Bogotá. Specifically, the aforementioned four-lane dedicated road for gasoline tax activities, quality assurance of crews, coordination with conventional classes, and thorough use of information and communication technology were not sufficient, and a comparison of the two cases shows a clear difference in quality. Nevertheless, the BRT system has gained a large number of users and a good reputation in the local community, and should be evaluated as a BRT case study in Southeast Asia.

5.4 REFORM OF BUS MANAGEMENT SYSTEM IN SEOUL, SOUTH KOREA

Fumihiko Nakamura and Yurie Toyama

5.4.1 Why Should We Learn the Experience in Seoul?

As an example of an Asian country that has learned from Bogotá's and Curitiba's experiences, it is appropriate to mention the case of Seoul, rather than the aforementioned Jakarta. Seoul is unique in that it is not a BRT project, but rather a remodeling of the entire bus system in the metropolitan area, and has achieved a radical transformation of the entire urban bus system in a short period of time.

All of the more than 8,000 buses, which were mostly privately operated, were placed under the control of the Seoul City Hall, which collected all fare revenues and distributed them according to the distance traveled, as in Curitiba and Bogotá. Vehicles were replaced one by one, repainted in four colors for regional trunk lines, city trunk lines, branch lines, and others, respectively, and system numbers were changed to a four-digit numbering system. Although there is no coordination in terms of city planning, it appears that the city has done everything it can as far as the transportation system is concerned.

5.4.2 Necessity and Background of BRT in Seoul

Seoul, the capital of the Republic of Korea, is a large city with an urban population of around 10,440,000 (2012). The city has undertaken various transportation restructuring initiatives, especially since 2000. For example, the removal of the motorway through the city center and the Cheonggyecheon restoration are well known.

In 2002, the Seoul Vision 2006 was established, which set out a major restructuring of public transportation with the aim of creating a human-centered environmental city. In particular, the bus system reformed in 2004 was a major innovation.

The population of the Greater Seoul metropolitan area tripled between 1960 and 2002, with a rapid increase in the number of private cars and traffic congestion. In 1974, the first metro line opened and began to serve as a mass transit system within the metropolitan area. However, with the prospect of further population and economic growth, there was a need for a transportation system that could transport people more cheaply and efficiently. This led to a reform of the bus network and fare system, as well as a review of the running space.

Prior to this, Seoul's bus routes were subject to various challenges such as private management, excessive competition, and dangerous driving, and service levels other than fares, such as timetables and routes, were not under the control of the government. Therefore, service levels were not always high, and citizens increasingly turned away from buses.

When Lee Myung-bak, who later became President of South Korea, ran for Mayor of Seoul, he was elected on a campaign promise to drastically improve the public transportation system in the city and to improve transportation problems by encouraging people to switch from using cars to public transportation.

The main points of the bus reform included placing bus companies under the control of Seoul City, reorganizing the bus route network into one consisting of trunk and branch lines, integrating the fare system and redistributing driving space by securing lanes exclusively for buses. At the same time, the operational methods were reformed, stops were modified, and information was reviewed (Pucher et al., 2005).

5.4.3 BRT Specifications in Seoul

The BRT system in Seoul consists of 12 corridors with 122 trunk lines (2014) and 215 branch lines (2014) with a total system length of 115.3 km (2014) and 329 stations (2014). Peak traffic is 8,400 passengers/hour/direction (2011), transporting 400,000 passengers per day (2009).

5.4.4 BRT Innovations in Seoul

5.4.4.1 Running Space

When restructuring the buses in Seoul, the Mayor of Curitiba, Federative Republic of Brazil, Jaime Lerner, was invited and the knowledge and lessons from Curitiba were also learnt.

When the buses in Seoul were restructured and partly converted to BRT, the driving space was also improved at the same time. Road space was redistributed and the center lane was made a dedicated bus travel lane.

The lanes were first introduced as dedicated lanes in the radial road network toward the center of Seoul, with the central part of the existing road space being used as an all-day dedicated lane (Figure 5.3). The width is also designed to allow overtaking at stops. The stops are designed as islands in the middle of the lane in conjunction with the central travel lane.

Figure 5.3 Bus lane in the middle of the vehicle lane

Source: (Photo taken by Yurie Toyama)

5.4.4.2 System Numbering Review and Livery

Another distinctive feature of the bus fleet is that the livery and system numbers were reviewed when the bus fleet was restructured. Existing bus routes, which had only two types – trunk and branch lines – were changed to four types – trunk, branch, circular, and wide-area buses – and the bus body color was painted according to the type of bus (Figure 5.4). In addition, the system numbers now represent eight zones, so as to judge from the system number alone where a bus is going from and to.

Whereas previously, each private bus company had designed its own body and system number on a voluntary basis, the city government introduced a management corporation to manage the private bus companies by introducing the concept of quasi-public management, with overall control by the city government. The role of this corporation is to manage routes and jointly purchase vehicles. Previously, private bus companies competed with each other for the most profitable routes, making it difficult to reorganize the routes as a whole. This systematic arrangement was possible.

5.4.4.3 Operational Management

All bus vehicles are managed by GPS, with boarding and alighting sensors installed at the doors, and the status of stops at bus stops is managed for each driver. The operation status is collated for each route, and bus standards are

Figure 5.4 Colored Bus at terminal

Source: (Photo taken by Yurie Toyama)

monitored and used to improve the service. Bus service status can be checked by users in real time at bus stops and by mobile phone. The frequency of bus services is managed for drivers, and the appropriate number of vehicles and timetables are managed according to demand.

In Seoul, this has led to IC cards also having a boarding rate close to 100%, allowing for more data analysis of usage results and usage characteristics.

5.4.5 Learning from the Experience in Seoul

The case of Seoul has distinctive features not found in Curitiba or Bogotá.

It did not install high-floor bus stations, in part because of the introduction of bus vehicles with low floor height; probably because of labor costs, it did not assign staff to bus stations. On the one hand, IC cards were introduced and spread at an astonishing speed. The cards can be purchased at convenience stores, and can be recharged. As of the 2013 hearing, more than 99% of the passengers were using the card. While there is an aspect of user service in the popularization of the cards, it is more likely that they are intended to increase speed by reducing stop times, although this seems to be the most important, to prevent fraud by bus crews by preventing crews from touching cash and eventually removing the fareboxes. It is important that IC cards allow city hall to manage all fare revenue flows.

Each bus stop also has a bus approach sign, and information is available on the Internet and later on smartphones. While this is also a service for users, it is more likely to be managed by the city office. Under a system in which contractual payments are made according to the distance traveled, the system is a device for ascertaining the actual distance traveled and for preventing the crew from canceling buses without permission.

In this way, it is clear that ICT technology was used even more than in Bogotá to avoid the risks associated with the transfer from conventional buses.

Also worth mentioning are two transportation policies that were advanced at the same time as the bus renovation. The first is on-street parking management on multi-lane roads. By adding parking where needed, the city has eliminated on-street parking on streets where buses travel. This ensures that even after the introduction of a center-traveled bus-only lane, there is still plenty of driving space for regular vehicles. The other is the enhancement of pedestrian space, represented by the Cheonggyecheon restoration and Namdaemun development on the east side of City Hall. With the intention of having buses handle the demand that cannot be met by subways alone, access to mainline bus stations and subway stations has become an important issue, and access to walking routes and adjacent public spaces has been secured.

Thus, Seoul's bus renovation was realized within the framework of a comprehensive urban transportation policy system, and it is a method that maximizes the use of ICT while incorporating the best aspects of Curitiba and Bogotá in a very strategic manner.

5.5 CONCLUSION

Fumihiko Nakamura

This chapter provides a detailed analysis of the introduction and operation of buses and BRT systems in urban transportation systems. It focuses on the current state of urban transportation in emerging economies and the development of BRT systems.

Buses play an important role in urban areas, especially in emerging markets, and are widely used as a means of transportation. The exact definition of a bus varies from country to country, but in general, a bus is a passenger vehicle of a certain capacity or more that is operated as part of the public transportation system, regardless of its mode of operation or vehicle type. In emerging markets, we can see different forms of bus operation in response to urban transportation demand.

One is what might be called community-based bus operations. In some cities, small bus services run by local residents are common. These usually use retrofitted vehicles and have emissions and safety issues, but they are an important means of transportation for local people.

A more common example is a service run by a local transportation authority. In these cities, bus services are provided by private operators under the control of the local government as a public transportation service. Although these bus routes are operated under a certain management system, they are not directly managed by the transportation authority and are often operated in a mixed way, with private companies usually cooperating in the operation.

Another form is large-scale operation by private companies. In some cities, private companies operate several bus lines, supported by commercial capital. Such operations often serve as an important means of low-cost transportation for low-income groups.

As urban areas expand and rail networks are underdeveloped, buses become the main mode of public transportation and are often used on a daily basis. However, the way they are operated is extremely diverse, and there are different needs and problems in each city. In addition, since many bus drivers' incomes are based on a percentage of fare revenue, there is a lack of income stability and safety may not be guaranteed (ADB, 2010).

BRT first attracted attention around 1996 as a transportation system that could provide faster mass transit than conventional bus systems. BRT aims to operate at high speeds by using dedicated bus lanes to avoid traffic congestion. The introduction of BRT systems has become popular in many emerging countries because they are less expensive and take less time to build than rail systems.

The origins of BRT can be traced back to the bus convoy system that began in the city of Porto Alegre in the 1980s and the main bus system in the city of Curitiba in the 1970s. These systems used buses that ran on dedicated roads to achieve high frequency and mass transit. The BRT system introduced in Curitiba in 1992 was innovative in terms of speed and efficiency and became a model for other cities.

The TransMilenio system, launched in 1999 in Bogotá, Colombia, became a model for BRT and influenced cities around the world. The Bogotá system uses dedicated roads and articulated buses, allowing for efficient operation and helping to solve urban traffic problems. This system was highly regarded around the world as a means of revolutionizing the image of conventional buses.

Bogotá's BRT system (TransMilenio) was developed based on the success of Curitiba. Bogotá introduced dedicated roads, high-floor bus stops, and articulated buses to improve the efficiency of bus operations. The management and operation of the system was separated between the operating company and the city government, with fare revenue collected by the city government and then distributed to the operating company.

The Bogotá system was introduced as a way to reconfigure the existing bus system and required a transportation strategy that was closely linked to urban planning. By widening the road in the middle as a dedicated road and introducing an express service that bypasses intermediate stations, operational efficiency was improved, and high-frequency buses supported citizen mobility.

Using the lessons learned from Curitiba, Bogotá operated the BRT with its own unique innovations. As a result, the city of Bogotá was able to reduce traffic congestion and redefine the value of buses as a public transportation system.

It should be emphasized that buses are and will continue to be an important mode of transportation in emerging markets, and that the introduction of well-planned BRT systems in large cities is bringing about important changes in urban transportation. BRT systems are being introduced in many cities as a low-cost, efficient means of transportation, and they are achieving success by working closely with urban planning. The cases of Bogotá and Curitiba show that BRT can contribute to urban transportation in a different way from conventional buses, but also that it is necessary to design systems that take into account the characteristics of each city.

REFERENCES

Asian Development Bank (ADB). (2010). Retrofitting sustainable urban development practices: Urban public transport. Author.

Cervero, R. (1997). *Paratransit in America*. Praeger.

Institute for Transportation and Development Policy (ITDP). (2007). *The BRT planning guide*. Author.

Nakamura, F., Makimura, K., & Toyama, Y. (2016). *BRT changes cities*. Institute of Behavior Sciences. (in Japanese)

Pucher, J., Park, H., Kim, M. H., & Song, J. (2005). Public transport reforms in Seoul: Innovations motivated by funding crisis. *Journal of Public Transportation*, *8*(5), 41–62. doi:10.5038/2375-0901.8.5.3

Yabe, T., Nakamura, F., & Okamura, T. (2006). A comparative analysis on the feature of planning and operation in the cities implemented bus rapid transit. *Journal of the City Planning Institute of Japan*, *41*(2), 8–17. doi:10.11361/journalcpij.41.2.8. (in Japanese)

Chapter 6

Railway

6.1 ADVANTAGES OF URBAN RAILWAYS IN EMERGING COUNTRIES

Toshiyuki Okamura

6.1.1 Introduction

Urban railways is a typical mass transit mode. In many developed countries, especially in megacities with a metropolitan population of over 10 million, urban railways plays an important role in urban transportation. For example, in Tokyo, which is currently the world's largest metropolitan area (metropolitan population of approximately 37 million in 2023), despite the fact that there are 330 passenger cars per 1000 people (2024), around 90% of trips from the suburbs to the city center are made by railways, and the share of railway for all trips in the whole metropolitan area is as high as 33% (52% if only motorized modes are counted; 2018). The urban railway is an essential element for this metropolis to function as a metropolis. Modern cities, not just megacities, need to ensure both the efficiency of socioeconomic activities and livable urban spaces and environments. If a city becomes heavily dependent on roads and cars for transportation, it will not only face problems of severe traffic congestion, air pollution (even though the EV ratio is expected to increase), and traffic accidents, but also be forced to reserve much of its urban space for cars due to road development and parking supply to mitigate these problems. Urban railways have various advantages, such as high capacity, high speed, time reliability, and environmental efficiency. Furthermore, in the cities of emerging countries, where it is expected that the ownership and use of cars are set to increase, there is a need for the development and utilization of urban railways, which have many advantages.

6.1.2 Features of Urban Railways and Their Benefits for Emerging Countries

The features of urban railways and their benefits for the cities of emerging countries can be described as follows:

DOI: 10.1201/9781003467403-7

6.1.2.1 High Capacity and Spatial Efficiency

Urban railways can transport around 80,000 passengers per hour in peak direction on a single track, according to the system. For example, the Marmaray commuter railway, which includes the Bosphorus Strait Tunnel section that opened in Istanbul in 2013, was designed to have a peak hourly transportation capacity of approximately 75,000 passengers per hour in peak direction (pphpd) at the time of planning (European Investment Bank, 2021). In 2019, just before the pandemic, some of Tokyo's suburban railways and subway lines were transporting around 80,000 passengers per hour (in 2023, after the pandemic, peak demand had decreased to around 65,000 pphpd). At this time, the passenger occupancy density was approximately 6 people per square meter of vehicle floor space, which is about the same as the density of an elevator that is practically full (not the capacity for safety regulations). (In addition, the commuter railway in Mumbai, India, which is said to be the "world's most crowded railway", is said to have 14 to 16 people per square meter of vehicle floor space at peak times [Pucher et al., 2004].) There is no other urban transportation mode that can transport as many passengers.

If this amount of transportation were to be handled by private cars, it would require a huge number of lanes and parking areas, and the city would not be viable in the first place. Even Bogotá's BRT, which has a very high transportation capacity even by world standards, can only transport 49,000 people per hour in one direction (2018) (WRI Brasil Ross Center for Sustainable Cities). Even that is with two lanes in each direction dedicated to the BRT. Even if the transportation capacity per hour is around 5,000 people the advantages of mass transportation offered by rail are still very much in evidence. To transport 5000 people by private car, assuming that each car carries 1.3 passengers, you would need around 3800 cars. Given that the traffic capacity per lane is around 1000 cars on city streets and around 2000 cars on access-controlled roads, to cope with this level of demand you would need either 8 lanes in each direction on city streets or 4 lanes on expressways.

6.1.2.2 Technical Expansion Potential

In the case of roads, the only way to deal with an increase in traffic is to increase the traffic capacity in proportion to the increase in traffic, that is, either by increasing the number of lanes or by building a new road. Even with BRT, once the capacity of the road is reached, it is almost impossible to increase it any further. Even in advanced BRT systems such as Curitiba and Bogotá, many years have passed since they began operating, and the issue of insufficient transportation capacity during peak periods has become a problem. The solution to this problem is no longer just a matter of increasing the number of vehicles or improving existing infrastructure.

In reality, the peak hourly capacity of many urban railways around the world is between 5,000 and 30,000 passengers at most. However, if the basic track infrastructure and space are used as they are, and the number of vehicles, signal equipment, and station facilities are increased, it is technically possible to increase the hourly capacity to around 80,000 passengers, as mentioned earlier. For example, over the 50 years from the 1950s to the 2000s, the transportation capacity of many suburban railway lines in Tokyo more than tripled (from 3–6 carriages to 8–10 carriages, and the frequency of operation also increased from about every 3.5 minutes to about every 2 minutes). In other words, urban railways are a system with high technical potential for future expansion. In particular, in large cities in emerging countries where future increases in transportation demand and transportation capacity shortages are expected, it is ideal to introduce mass transit (i.e., urban railways) with high potential for future expansion from the relatively early stages of urban growth and economic development.

However, the future expandability of a system depends greatly on the vehicle specifications (especially the cross-sectional size) and the basic track space size, and in the case of underground railways, the expandability of the platforms is critical. In the case of introducing small-sized vehicles or somewhat specialized infrastructure such as monorails or Auto Guided Transit (AGT), the future transportation capacity is strongly dependent on the initial infrastructure level. These modes should be considered as medium-capacity transportation systems, and it is preferable to introduce them in complementary corridors rather than in the main public transportation corridors within the city. In addition, even if it is technically possible to upgrade them, in many cases, particularly in emerging countries, it is difficult to procure the necessary funds from a financial perspective, so upgrading and improvement do not proceed. These are discussed later.

6.1.2.3 Travel Speed and Time Reliability

Urban railways that can operate independently of road traffic have high travel speed and time reliability (punctuality). This is their greatest competitive advantage over cars. These two factors are the most effective in persuading people to curb their car ownership and use. In the case of a metro system in a central urban area, the average travel speed, including dwell time at each station, is largely dependent on the interval between stations. If the average station interval is 1 km or less, the average speed is around 30 km/h. In the case of private cars, even when there is no traffic congestion, the average speed in central urban areas is around 20km/h, and when there is congestion, it drops significantly lower. Even with BRT, which has dedicated lanes, the average speed in central urban areas is around 20 km/h. This difference is more noticeable to users as a shorter travel time when the distance travelled is longer. Furthermore, in the case of suburban railways, when the interval between stations is longer or when express trains are in operation, the average speed is between 40 km/h and 70 km/h. This is almost equivalent to a highway with no traffic congestion. In the cities of emerging countries, which are continuing to expand, the distance of travel across the metropolitan area (such as commuting, business, shopping, and others) is increasing due to the expansion of the urban area, and the distance of travel within the city center (business, shopping, and others) is also increasing due to the expansion of the city center itself and the development of sub-centers. In the cities of emerging countries, where economic development and the sophistication of business are progressing, the mass transportation capacity of urban railways and their high travel speed and punctuality are important.

6.1.2.4 Environmental Efficiency and Urban Development Capabilities

Railway systems, especially those using iron rails and wheels, are extremely energy efficient per passenger transported. This characteristic is even more pronounced when the transportation density is high. Many urban railways are electrically operated, and the motors are highly energy efficient and do not emit air pollutants when in operation. The electrification of automobiles is also progressing rapidly, and the major advantages of this are that they do not emit air pollutants when in operation and are quiet. However, in terms of energy efficiency, the latter is still overwhelmingly higher than that of cars.

From the perspective of urban space, in addition to the space efficiency from the perspective of transportation efficiency mentioned in Section 6.1.2.1, the small space of dedicated tracks means that space can be allocated for urban and environmental functions. Urban railways have high space efficiency at stations and transportation nodes. Therefore, in city centers and sub-centers, it is possible to create spaces that combine high levels of urban concentration and urban amenity by linking the development of urban transportation with the concentration of urban functions around stations, and by further ensuring the provision of pedestrian spaces and open spaces around stations. In residential areas, it is possible to achieve a location and people's behavior characteristics that enable a lifestyle within the station sphere that is centered on walking and does not rely on cars. This kind of Transit-Oriented Development (TOD) is what urban railways are best at, and it brings out the potential of both the city and the urban railway.

Although TOD is a concept that is applied to new development areas in a narrow sense, the competitive advantages of urban railway systems over private transportation, such as their ability to transport large numbers of people, their space efficiency, and their high speed, mean that "TOD-style" urban areas and lifestyles will gradually form naturally in existing urban areas along urban rail lines and in unplanned urban areas. Although this is not often pointed out, in emerging countries where not all urban development is necessarily "planned", I would like to point out the possibility of environmentally sustainable urban development being formed spontaneously through urban railways as a latent benefit of urban railways in emerging countries.

6.2 TYPE OF URBAN RAILWAY AND ITS DEVELOPMENT

Toshiyuki Okamura and Fumihiko Nakamura

6.2.1 Types of Urban Railways

Up to this point, the term "urban railway" has been used without any particular definition. There are various modes of urban railways, and the names and terms used to describe them are not necessarily strictly defined. With this in mind, we will divide them into several types. Typical modes of urban railways are as follows:

6.2.1.1 Metro, Subway, or MRT

These are modern urban railways newly introduced in the city, separate from conventional railways, with an exclusive right-of-way independent of road traffic, and the network is basically closed within the metropolitan area. It is positioned as Rapid Transit (RT) or Mass Rapid Transit (MRT) in the sense that it is a rapid railway that differs from trams, which share road traffic space. Some cities in emerging economies that have newly introduced urban railways use the term MRT as the name of these rail systems. Historically, Metro developed in cities in developed countries as a replacement for the role of intra-urban trams and buses, and has developed as a rail system for intra-urban transportation in cities. They are often built underground in already urbanized cities, so that in some cities, especially in developed countries, the Metro is almost synonymous with subway railways. However, from a global perspective, whether a line is underground or overground is essentially irrelevant, and there are many sections that are built above ground (often in elevated structures). In some cities, the Metro extends from the city center to the suburbs, providing transportation not only within the city center but also between the suburbs and the city center. In other cities, commuter rail/suburban railways (including conventional railways) provide transportation between the suburbs and the city center. The standards of rolling stocks vary from those with the same or almost the same capacity as the heavy rail type of intercity rail, to those with medium-sized rolling stocks of slightly smaller width and cross section (including those running on rubber tires), and those with rolling stock standards almost equivalent to those of Tram/Streetcars. In the case of subway railways, a standard with a slightly smaller cross section is often selected from the viewpoint of reducing construction costs. In such cases, however, the transportation capacity is also small, so it is important to determine the vehicle standards in advance according to the expected future demand.

6.2.1.2 Automated Guideway Transit (AGT), Monorail

AGT is an automated guideway transit system operated by rubber tires and so on, while a monorail is a single-track transit system with either suspension or straddle-type systems. The key points of these systems are that they can handle steep gradients and sharp curves, have a smaller vehicle cross-section and track width, and save on construction and land acquisition costs compared to standard Metro/MRT systems. They also have an exclusive right-of-way track space independent of road traffic, and are mainly constructed as elevated structures. In this sense, they are categorized as Metro/MRT. Although the initial cost of construction is low, which is why it is sometimes chosen in emerging countries, the cost of rolling stocks and operating costs (vehicle maintenance and power costs) per unit of capacity are relatively high. Since the level of infrastructure was originally designed to minimize initial costs, if demand increases more than expected later, the cost of infrastructure expansion is often relatively high or expansion itself is difficult to achieve. It should be understood as an effective mode for medium-volume transportation with small transportation capacity in cases where there is no unexpected increase in demand in the future.

6.2.1.3 Light Rail Transit (LRT)

LRT is a modernized system of vehicles and tracks based on the Tram/Streetcar vehicle standards for on-road operations. As a vehicle type, it is classified as light rail. LRTs can be divided into two types: those that run on roads (modernized Tram system) and those that run on exclusive right-of-way. While the former does not differ from BRT in terms of average travel speed and capacity, it is more expensive to implement, and there are relatively few examples of its use in emerging countries. The latter are classified as medium-volume transit in the Metro/MRT category. Some cities refer to Metro/MRT with medium-sized vehicles as LRT, and it is often chosen in emerging countries because of its lower initial cost compared to heavy rail. It should be noted that this mode also has somewhat lower potential for future expansion.

6.2.1.4 Commuter Rail/Suburban Rail

These are cases where an intercity rail line is also operated as an urban rail line, or an existing rail line is modernized to become an urban railway. Because of this, in developed countries, most of these lines provide service between suburban areas and urban centers (or the fringes of urban centers), and are naturally distinguished from metro lines, which mainly cover inner-city transportation. In many cases, the operator is a state-owned railway (or its privatized operator) that operates intercity services, which is different from the abovementioned Metro. However, when Metro/MRT by heavy rail uses the same rolling stocks and system standards as suburban rail/commuter rail (e.g., Tokyo, Seoul), the technical distinction between the two is not very meaningful. In emerging economies, there are many examples of existing intercity railways within cities that have great potential for use as urban railways, but have been slow to modernize. In addition, there are cases where the linkage with Metro/MRT is insufficient and the two are not well formed as an integrated network.

6.2.2 Formation of Urban Railways Network

Urban railways must function as a network, rather than as individual lines existing as stand-alone entities. When lines intersect, there must be a plan for routes and nodes that are easy to transfer between, and there must also be a division of roles with BRT and bus systems that serve as feeders. To this end, a master plan for urban railways must be developed, whether in developed or emerging countries, not only to create a mass transit network for the city as a whole, but also to ensure that individual lines are efficiently interconnected, taking into account their consistency as a network. In doing so, the choice of technology (type of rail) to be introduced on individual lines is also important. The following describes the stages and types of urban railways network formation, assuming cities in emerging countries.

6.2.2.1 Stage of Introducing Metro/MRT for the First Time

As mentioned earlier, when cities in developed countries started to develop Metro in the past, the suburban rails already existed as urban railways, and Metro's main role was to carry out intra-city transportation. In contrast, cities in emerging countries often do not have existing urban railways or have inadequately developed them, so Metro must also be responsible for travel along the main corridors connecting suburbs and urban centers. In addition, routes are often selected in areas where a large number of users can be expected and where it is easy to construct new lines (i.e., areas where existing wide streets or rivers can be used to construct elevated or underground structures, or where land acquisition is easy with urbanization in progress), and construction is often started on those sections.

In the early stages of development, the most common routes that are expected to attract many users are: 1) routes that connect the city center and subcenters (routes that serve major corridors in inner city areas) and 2) routes that extend from the city center to the suburbs (including routes with both ends located on the suburban side). Then, while progressively extending toward the suburbs, a second and subsequent line extending in the direction of another suburb will open in the same sequence.

Especially in the case of using the space of existing arterial roads, which are already important urban transportation corridors with populations and various facilities located along them, a certain number of users can be expected simply by shifting from another mode of transportation. Furthermore, the high accessibility of both arterial roads and urban railways will greatly encourage new locations, creating a positive spiral that will further increase the number of rail users. For example, Bangkok's urban railways, the Skytrain (Bangkok Transit System: BTS), was built on an elevated track above the existing arterial street, and since its initial opening (1999), it has been extended in a route pattern that is exactly the combination of 1) and 2) above. The line has been increasing the number of users as it penetrates through the central business and commercial districts, and new locations have been promoted along the line. However, when urban railways are built on arterial roads radiating from the city center, there are cases where the opening of the section on the city center side is delayed due to delays in land acquisition or construction of underground structures, or where the number of users is sluggish in the early stages of operation due to the poor location of the terminal on the city center side. In such cases, the feeder transit system on the city center side should be enhanced in the short term, and in the medium term, the network effects of the opening of the city center side and the opening of other lines should be awaited. Considering the profitability of urban railways, the slow growth in the number of passengers in the early stages of opening should be avoided as much as possible, and the selection of the sections and terminal locations on the city center side is important in the initial introduction stage of urban railways in emerging cities.

6.2.2.2 Dealing with Insufficient Capacity after Opening

The choice of the mode of the railway to be introduced is important. Especially in emerging countries, it is important from the perspective of business continuity to satisfy both initial cost reduction and ensuring the number of users in the initial phase of the opening. As mentioned in Section 6.1.2, many routes that introduced small- and medium-sized rolling stocks in order to reduce construction costs often suffer from capacity shortages after their launch. For example, Manila's urban railways, LRT Line 1 (opened in 1984) and MRT Line 3 (opened in 1999), are both elevated railways with small

Figure 6.1 Congestion of LRT Line 1 in Manila

Source: (Photo taken by Toshiyuki Okamura)

LRT-type vehicles, and since their routes are located on major arterial roads in the city, they have a very large number of passengers. However, lack of capacity has been a long-standing issue. The difficulty in solving this problem lies partly in the fact that fares are inexpensive and the system relies on operating subsidies, which poses financial problems, but essentially it is the small size of the rolling stocks and high cost per capita of passengers transported that make it difficult to radically increase its capacity as an infrastructure.

In cases where future demand growth is anticipated from the planning stage, larger standard rolling stocks are often introduced. In the early years, Metro in Seoul (1970s), and in the 1990s and later, mega-cities in emerging Asian countries such as Beijing, Shanghai, Bangkok, Delhi, Dhaka, Jakarta, and Ho Chi Minh City, have adopted large-sized vehicles. In some cities, trains are introduced in short units at the time of opening, and the infrastructure is designed to allow for additional vehicles in the future. However, there are some cases where the number of rolling stock cannot be increased despite the increase in the number of passengers, not due to technical constraints, but due to operating methods and financial constraints.

6.2.2.3 *Stage of Implementing Multiple Routes*

If a master plan has been formulated in advance, the phase of development is also considered, and the station structure is designed to facilitate transit access at the points where lines will intersect in the future. In reality, however, projected population growth and traffic demand growth, as well as economic conditions, often differ from the original assumptions, often delaying the realization of the original master plan or forcing changes to the plan. In addition, even if a master plan exists, it is often limited to a macro route plan, and the network is expanded sequentially without a detailed plan for transfer-friendly nodes, resulting in a structure and layout that makes it very inconvenient to transfer at the additional transfer stations to be built.

In addition, different lines have different financing methods and therefore different operating entities. This can lead to different fare structures, and in some cases, it is difficult to improve the infrastructure for transferring between lines. The most appropriate rail mode and standards are to be determined for each route, but the overall operation may become inefficient because of the incompatibility of multiple systems, each of which is incompatible with the other.

At the planning stage of implementing multiple lines, it is important to consider the consistency of the network as a whole at the planning level, to take into account the ease of connections at nodes in advance, and to examine the advantages and disadvantages of mixing multiple systems. At the operational level, when different operators operate different routes, it is important to coordinate with each other (fares, information, and signage at nodes). It is also important to design the scheme in such a way that the different operating modes do not interfere with this coordination.

6.2.2.4 Cases with Existing Rail Infrastructure

Many large cities in emerging countries had railways built before World War II, primarily for intercity passenger and freight transportation. Although some of this rail infrastructure still exists today and is used for urban transportation, it is often in a poor state of maintenance. In many cases, the rolling stock consists of old passenger cars and locomotives. In addition, the managing and operating bodies are often state-owned enterprises directly linked to the government, and most of them are in a poor state of management. In post-war urban development, it did not begin around railway stations, but rather along newly constructed or widened roads. In particular, middle- and high-income new residents prefer to live along arterial roads. As a result, the areas around stations of conventional railways became relatively low-income neighborhoods and were positioned outside of the city's overall structural system. Despite the availability of such conventional railways in large urban areas, conventional railways have rarely been emphasized in urban transportation planning for the reasons mentioned earlier. Even in urban transportation projects supported by developed countries, elevated railways and Metro were often proposed without mentioning conventional railways.

Bangkok, with its expanding urban railways network, also has conventional railways that serve intercity transportation. Although improvements have been made compared to the past, local trains, in particular, are operated by old-style passenger cars and short diesel trains, and the track, signaling, and other facilities are also outdated. The MRT and the Airport Link are constructed by elevated tracks, completely separate from the conventional railways, using the same land along the surface tracks. Conventional railways are not included in the master plan for urban railways, and fares are set at about one-tenth the cost of MRT. The old and new rail systems operate in parallel, without any interconnection. Cases of old-style conventional railways (including suburban railways) and modern Metro/MRT systems coexisting without any coordination can be seen in cities in emerging countries such as India, Brazil, and South Africa. More consideration should be given to utilizing and upgrading conventional railways into urban railways.

There are many examples of conventional railways with this great potential being converted into modern urban railways through rehabilitation and upgrading. In the Kuala Lumpur metropolitan area, commuter service is being provided through the modernization of Malayan Railways, along with MRT and other improvements under the master plan for urban railways. The commuter rail in Istanbul, called Marmaray, has a total length of 76.3 km with an average daily ridership of approximately 600,000 passengers (2024). During the daytime, a 10-car train operates every 5 to 10 minutes. Commuter rail/suburban rail in Jakarta, Indonesia, was electrified and tracked as urban railways in the 1970s, but the level of maintenance of rolling stock was low, while extremely crowded during peak hours and people riding on the roofs of trains became a common practice. In the 2010s, the operator began full-fledged modernization of its facilities and improvement of station facilities, and introduced more than 1,000 trains formerly used in Tokyo to increase transportation capacity and improve safety and comfort, while training engineers and crew members. In Manila, a project is underway (as of 2024) to substantially upgrade the non-electrified lines of the Philippine National Railway to commuter rail. The utilization of these conventional railways and the conversion to commuter rail will require not only technical modernization, but also coordination and integration with the new urban railways, the MRT.

In converting conventional railways to commuter rail, one option would be to specialize in the demand for commuters without high-frequency operation, and to operate a few trains in the direction of the city center in the morning and a few trains to the suburbs in the evening (this type of railways also exists in North America). For example, in Nairobi, Kenya, national railway (Kenya Railways) provides commuter service as described earlier. The coaches used for long-distance trains have been converted to bench seats for commuters, and capacity has been secured by longer trains, though only a few round trips per day, and at the same time, stations have been renovated to enhance security. A feeder bus service from the central terminal station to and from the central business district operates in conjunction with the train service. Currently, trains are mostly occupied with passengers being able to be seated, and comfort is ensured. In Nairobi, which relies on informal paratransit for urban public transportation and has extremely heavy road congestion during peak hours, the commuter rail has received some recognition for its punctuality, comfort, and safety. The redevelopment of conventional railways through such relatively low-cost investments is one way.

A circular rail line in the Yangon Metropolitan Region in Myanmar is one of the examples that have decided to rethink conventional railway in response to urban development. This line was planned, designed, built, and opened in 1959 by British and German engineers. There are no railroad crossings in downtown Yangon, and all major roads were designed from the beginning to be multilevel. At the time, no other urban rail line in Asia, including developed countries such as Japan, had such a thoroughgoing system. Due to subsequent changes in the political situation, the ring railway was largely neglected, and the urban transportation plan for the Yangon metropolitan area by the Japanese government and others did not include

Figure 6.2 Circular Railway in Yangon

Source: (Photo taken by Fumihiko Nakamura)

the use of conventional railways, but only plans for elevated railways. Since 2009, however, the rehabilitation of the railways has been promoted, and support measures have begun, including not only track repairs and improved locomotive maintenance, but also the introduction of second-hand Japanese diesel railcars, resulting in more frequent and faster service than in the past. The number of users has begun to increase, and the service is not necessarily limited to the poorest of the poor. However, the land use around the stations is not planned at all. Although the stations are located near airports, large industrial parks, and new residential areas, there are still problems in securing the flow lines between the stations and the land. Nevertheless, it is significant that a case has been made for the use of conventional rail.

The situation is different in other countries; however, the planning option of promoting conventional rail infrastructure in conjunction with station area redevelopment utilizing unused land of railway operators appears to be a viable option.

6.3 FINANCIAL MANAGEMENT AND OPERATING BODY OF RAILWAY

Toshiyuki Okamura

6.3.1 Costs of Urban Railways

The initial costs (mostly fixed costs) of urban railways are high. The price of a single rail car (about 20 m in length) is roughly US$ 1 million (excluding maintenance contracts), and construction costs are typically US$ 100 million per km for an elevated (double track) structure and US$ 200 million or more for an underground structure. Furthermore, a long period of time is required from the decision to commercialize the project to the design, construction, and opening of the line, during which time risks such as increased costs and changes in economic conditions may arise. In addition, operating and maintenance costs (often variable costs) after opening are also significant.

Although there is a wide range of procurement costs depending on local conditions, there are no major differences on a global scale except for costs related to local labor costs. However, especially in emerging economies, the level of affordable fares for people is highly dependent on the economic level of the city (e.g., GDP per capita). Even in developed countries, it is difficult to cover all the fixed and variable costs of urban railways directly from users' fare revenues, but at least a certain percentage of variable costs must be covered by fare revenues to ensure business viability. Therefore, even if one tries to set affordable fare levels, urban railways fares must be set at a premium in relatively low-income countries. Low per capita costs (i.e., high ridership) are necessary to keep fare levels in check. Therefore, on the one hand, from the perspective of the business feasibility of urban railways in relatively low-income countries, the construction of urban railways will be limited to the trunk corridor sections in large cities where a larger number of users can be expected, and where the demand for such services is particularly high. On the other hand, from the perspective of urban transportation planning, as mentioned earlier, it is ideal to introduce urban railways at a relatively early stage of urban expansion and economic growth. The difficulty of developing urban railways in emerging countries lies in the fact that these two contradictory requirements must be satisfied.

For example, the fare for a 5 km section of urban railways is US$ 0.41 (3 RMB) for the subway in Beijing (China), US$ 0.54 (2.4 Ringgit) for the LRT in Kuala Lumpur (Malaysia), US$ 0.37 (6000 Rupiah) for the MRT in Jakarta (Indonesia), and US$ 0.23 (Rs. 20) for Delhi Metro (India; in 2024). The per capita GDPs of China, Malaysia, Indonesia, and India are US$ 12720, US$ 11993, US$ 4788, and US$ 2410, respectively (World Bank Group, 2024). While taking into account the ability of citizens to pay for urban railways, the fare level cannot be reduced to half even if per capita GDP is halved. The fare level that can cover some of the operating costs of urban railways is no less than US$ 0.2 per 5 km, and that is when there is a significant number of expected passengers. From this perspective, cities in emerging countries where new urban railways can be implemented are mega-cities (population of 10 million or more) in countries with a per capita GDP of US$ 2,000 or more. As the income level rises, the network can be expanded and urban railways can be introduced to more medium-sized cities. However, since it takes at least 10 years from conception to planning, construction, and opening of urban railways, the conception and planning of urban railways should start when GDP per capita is less than US$ 2,000, assuming that the aforementioned fare levels can be set at the time of opening of the railways.

6.3.2 Financing and Operational Structure

How to secure the initial costs required for the development of urban railways is a critical issue in emerging countries, especially in lower-middle-income countries. As in the case of road development, in addition to public funds, there are other sources such as ODA and loans from Multilateral Development Banks (MDBs) such as the World Bank and the Asian Development Bank (ADB); PPP may also be used to leverage private capital. In the case of urban railways, it is more difficult to secure revenue from users than from roads because of the need to make the fare level affordable. In addition, it is difficult to forecast demand, especially in the early stages of operation, and it is often not possible to secure the projected demand. In addition, costs include not only civil engineering work but also electrical equipment, station equipment, and rolling stock facilities, and the risk of project delays is generally greater than for roads. Furthermore, the initial costs must be reimbursed, as well as the operating costs and revenues must be balanced. For these reasons, urban railways are more challenging from a business feasibility perspective than roads, airports, or ports. PPP frameworks such as Build-Operate-Transfer (BOT) and Build-Transfer-Operate (BTO) are often applied in infrastructure projects. In the past, due to the aforementioned characteristics of urban railways in emerging countries, there have been many cases where the constraints of PPP frameworks have stalled operational improvements or additional investments, or where the PPP frameworks have failed and the government has bailed out the private operator due to the constraints of the PPP framework. In urban railways, it is desirable to avoid excessive risk-sharing to the private sector by utilizing public funds or ODA rather than private funds for the initial costs (especially infrastructure costs).

In railways, an operating structure that separates the ownership of the infrastructure section from the operation of the trains is often adopted. For operating bodies, it is desirable to utilize the management flexibility and know-how of the private sector. There are many cases where different operators operate different routes within the same city. The integration of operating bodies, such as the Transit Alliance in Europe, is not necessarily the best and only ideal operating system for urban public transportation in all emerging countries. However, at the very least, it is essential to coordinate the soft aspects, such as common fare payment methods, integration, and common use of transit information and maps, and the hard aspects, such as easy-transfer station infrastructure.

6.4 COORDINATION OF RAILWAY SERVICE WITH OTHER TRANSPORTATION MODES

Toshiyuki Okamura

Urban railways are not the only means of transportation within a city. It is also impossible to cover the entire urban area by urban railways alone. Especially in cities in emerging economies, maximizing the use of urban railways as a whole, which is an expensive and regionally limited mode, will contribute to improving the viability of urban railways, strengthening urban transportation systems, and contributing to the sustainable development of cities. In order to maximize the benefits of urban railways, it is important to coordinate not only among the aforementioned rail lines, but also with other urban transportation modes.

6.4.1 Improving Accessibility to Stations on Foot

Although often forgotten, the accessibility of non-motorized modes, mainly walking to stations, is extremely important for improving urban railways and urban mobility. The fragmentation of walking access by car-centered roads around stations is a major obstacle to the use of railways. In addition, the development of pedestrian space within walking distance of stations should be addressed in conjunction with the development of railways.

6.4.2 Bus and Paratransit Connections

In both the suburban and urban areas, it is important to improve accessibility to rail by providing bus transit facilities alongside rail stations. Nevertheless, for bus operators, it is generally more advantageous from a business standpoint to have passengers ride directly on bus routes parallel to railways without having to transfer from the bus to the railways. If a hierarchical network with railways as trunk lines and buses as feeders is desirable from the perspective of urban transportation planning, it is necessary to reorganize bus routes to accommodate the extension of railways. Even in cities where such coordination is institutionally difficult, if urban railways are overwhelmingly competitive in terms of travel time and time reliability, the bus network will naturally be adjusted to a hierarchical network by the operators. In the case of informal public transit (paratransit), which is extremely market-competitive and adapts to changes in demand, supply will spontaneously originate in front of stations. Boarding and alighting on the streets around stations not only becomes a bottleneck for traffic flow, but also leads to road safety issues and makes it difficult for users to understand. It is desirable to regard not only formal buses but also informal paratransit as rail feeders, and to plan off-street transfer facilities for this purpose in advance.

6.4.3 Pricing of Fares

Common fare payment methods (e.g., IC cards) are becoming more common in emerging countries. However, there are few examples of common fare systems in public transportation within metropolitan areas in emerging countries, and there are many obstacles to its realization. If fares for each mode of transportation are at an affordable level, users' resistance will not be so significant, even if they have to pay a fare for each transfer. A fare system that causes significant user resistance to transfers is, for example, one in which each route or operator has a flat fare regardless of distance. For short-distance routes, the flat rate is not so high, but if long-distance routes extending to the suburbs have flat fares, users will feel great resistance to transferring to different routes or modes. The Istanbul Metro has flat fares for each line, even when operated by the same operator, and many Latin American cities have flat fares for intra-city railways and buses (resulting in a higher fare for short-distance use). These are due to a variety of reasons, including past history, and cannot be said to be generally problematic. However, in order to increase the convenience of the network through transfers and to encourage the use of public transportation for short-distance travel, excessive flat fares by route may be undesirable. The technical constraints to introduce a distance-based fare system with a common means of payment and less resistance to transferring are becoming smaller.

6.4.4 Relationship with Private Transportation

There are cases where park-and-ride facilities are attached to stations. At stations on the suburban side, when roads parallel to railways are congested, the advantage of transferring from private cars to railways arises. The usage is highly dependent on pricing. If the park-and-ride is operated by a public authority or a rail operator, it is important to set fees strategically according to policy objectives (e.g., maximizing profits in the parking business as a profitable business, reducing road traffic, improving accessibility on the suburban side). It should also be noted that setting a lower rate for park-and-ride will compete with feeder use by public transportation (resulting in a decrease in public transportation use and an increase in automobile traffic at the feeder portion). Furthermore, it should be noted that since the area surrounding the station is a predominantly residential location and has high potential as a commercial and other hub, a park-and-ride parking facility directly connected to the station would compete with those uses. Considering these factors, from an urban transportation planning perspective, park-and-ride facilities should be limited to cases where the competition with feeder public transportation and urban land uses is small and the benefits of transferring from automobiles are large (e.g., stations near highway interchanges).

It is also desirable to provide parking on a small scale and at a slightly higher rate (equivalent to the market price) as a possible alternative for car users. In contrast, park-and-ride by motorcycle is beneficial for both users and railways, since the parking space per vehicle is much smaller than that of a car, and railways have a higher advantage in terms of speed and comfort.

6.5 CONCLUSION

Toshiyuki Okamura

Urban railways – with their capacity as mass transit, technical expandability, higher travel speeds, and time reliability – environmental efficiency, and urban development capabilities are indispensable for cities in emerging countries where urban populations are concentrated in cities and urban areas continue to expand. Ideally, urban railways should be actively introduced from the relatively early stages of urban expansion, economic growth, and increased car ownership in order to build stronger urban transportation systems and achieve sustainable urban development. New urban railways in emerging countries need to be able to appeal to both the relative middle and lower income groups that do not (or cannot) own private transportation, as well as to each of the middle and upper income groups that do (or potentially could) own private transportation. In other words, fares must be affordable and competitive with private transportation (cars and motorcycles) in terms of travel time and comfort. However, the initial cost of urban railways is high, and it is difficult to finance their introduction and provide affordable fares when the income level of the cities is still low.

From the standpoint of the business feasibility of urban railways, they have no choice but to start their development only on trunk corridor sections in large cities where a larger number of users can be expected, and where the demand for such services is particularly high. Furthermore, it is necessary to control initial installation costs by, for example, reducing the size of vehicles and curbing the scale of infrastructure. However, in the medium term, the potential for urban transportation demand growth in emerging countries is large, and in many cases, the initial level of infrastructure will be insufficient to meet the transportation capacity needs. Thus, cities in emerging economies face the difficulty of developing urban railways while satisfying a number of conflicting conditions.

A frequently asked question in emerging countries is "Which should be selected, MRT/LRT or BRT?"

If BRT were cheaper than rail in all cases, and could provide transportation capacity and reliability nearly equal to that of rail, the answer would be simple. However, the advantages and disadvantages of urban railways and BRT are clear in this chapter. Longer travel distances within a metropolitan area favor urban railways because of their faster travel speeds. BRT occupies more of the road space on the ground, so the spatial efficiency of urban railways is higher. In order to achieve compatibility between the efficiency of socioeconomic activities and high urban density with livable urban space and urban environment, urban railways are advantageous because they require less surface space for rolling stock and terminal space, and have higher transportation capacity. Although the initial cost of urban railways is high, the average cost per user will drop significantly if the number of users of the entire system is high. In the case of BRT, the average cost per user will drop if the ridership per each bus is high, but the average cost per user of the entire system will not drop so much if the number of users of the entire system becomes high.

Furthermore, urban railways are more responsive to significant increases in demand over the medium to long term because of their high technological expandability. In cities with large populations and large urban areas, cities with dense central business and commercial districts, and cities where significant future increases in transportation demand are expected in the medium to long term, it may be desirable to have urban railways as the primary urban public transportation system and BRT as a more regional public transportation system or as a feeder to urban railways. The BRT will be able to play such a role in the medium to long term. When comparing such a role between LRT and BRT, which have smaller transportation capacities, there is no significant difference in transportation capacity and speed, so the decision should be based on other factors (environmental performance, spatial characteristics) and the specific characteristics of the areas where they are planned to be introduced.

REFERENCES

European Investment Bank. (2021, December). *Project completion report: Bosphorus Tunnel & Bosphorus Tunnel Tranche B Turkey*. Retrieved from https://www.eib.org/attachments/pipeline/20130129-pcr-public-use.pdf

Pucher, J., Korattyswaroopam, N., & Ittyerah, N. (2004). The crisis of public transport in India: Overwhelming needs but limited resources. *Journal of Public Transportation*, 7(4), 1–20. doi:10.5038/2375-0901.7.4.1

World Bank Group. (2024). GDP per capita (current US$). *World Bank Open Data*. Retrieved from https://data.worldbank.org/indicator/NY.GDP.PCAP.CD

WRI Brasil Ross Center for Sustainable Cities. (n.d.). Peak load (passengers per hour per direction). *GLOBAL BRT Data*. Retrieved from https://brtdata.org/indicators/systems/peak_load_passengers_per_hour_per_direction

Part II

Transportation Management

Chapter 7

New Services Aided by ICT

7.1 OVERVIEW OF THE EMERGENCE OF NEW IDEAS

Yurie Toyama

The introduction of information and communication technology (ICT) measures and services for urban transportation has become increasingly widespread in emerging countries in recent years. According to the Ministry of Economy, Trade and Industry's White Paper on Trade, Japan (2020), a country-by-country comparison of the number of registered cell phones per 100 people shows that Japan has 139, South Korea 130, and the US 124, while Thailand has 180 and Vietnam 147, indicating the tremendous penetration of cell phones in the Southeast Asia (SEA). Although developed countries were ahead of the rest of the world in terms of access to telephone lines and internet access, with fixed-line telephones and other equipment having been developed ahead of the rest of the world, the fact that cell phones are less costly to develop infrastructure than fixed-line communications led to the spread of cell phones at once, without going through the process of fixed-line communications (so-called leapfrog).

Thus, urban transportation services based on smartphone are being developed, especially in the SEA region, where more than one smartphone per capita is widely used.

The benefits of using ICT in emerging regions are many. From the user's perspective, the ease of gathering information has increased dramatically, the user side is closer to having complete information, and the user is able to know and choose among the mobility options available. Outside of the track-based transportation network such as train and metro, buses, paratransit, and other local services that are "if you know it, you can ride it," there is a large disparity in the ease with which people can get around between knowing and not knowing. In emerging countries, there were many situations where users were forced to make choices under conditions where they did not have complete information on what means of transportation were available from the origin to the destination, what kind of timetable they had, how much

DOI: 10.1201/9781003467403-9

the fare was, and what kind of transportation operator operated the service. However, advances in ICT means have expanded the potential for users to look up routes, timetables, and fares for transportation services with a smartphone at hand, and to make payments as well. Often, transportation services in emerging countries were not well informed or there was no way for users to know their options as complete information due to differences among transportation operators, challenges in the way transportation operators operate, and other factors. For example, users were not able to find out how to transfer between different operators, or to get suggestions for alternative means of transportation. Also, in the case of privately operated tuk-tuks and motorcycle cabs, the fare was negotiated on a case-by-case basis, making it impossible for passengers to know the market price of the fare. However, with the spread of smartphone applications, it is now possible to search routes on a map, negotiate the best fare according to the supply-demand balance, and estimate the time required. Without ICT technology, it would not have been possible to travel safely in the often heavily congested road traffic, where it is difficult to read the time required when using transportation services that run on non-exclusive routes, or to keep track of driver information.

However, ICT technology has only expanded the possibilities as a means of realization, and in reality, there are still many organizational and ecosystem challenges, such as the hurdle of understanding the actual operation of each means of transportation, its routes, fares, availability of timetables, and so on, and the hurdle of converting such information into digital data. In fact, many organizational and ecosystem challenges remain, such as the hurdle of understanding the actual status of each means of transportation, the hurdle of converting such information into digital data, the hurdle of continuously updating such information, and so on.

There are, of course, benefits for operators and policy makers just to overcome these challenges. Rather, without the need to build a large-scale infrastructure, the advantage of smartphone penetration can be used to collect travel data for policy making and better service, to understand the status of operations, and to manage vehicles and personnel for safer and more efficient operations. Providing safe transportation services will also help reduce traffic accidents and deter in-vehicle crime, including sexual assaults. Furthermore, in emerging countries, where it has been difficult to collect data on actual movement, the use of payment data and other data will lead to the acquisition of OD data on people's movement.

One of the reasons behind the push to popularize transportation and urban services using new ICT technologies is the promotion of smart city policies in recent years, particularly in SEA countries. For example, Thailand established "Thailand 4.0" as part of its 20-year national strategy (Irvine et al., 2022). Thailand 4.0 is a vision for sustainable economic growth, which includes smart city development and is promoted by the Thai Digital Economy Promotion Agency (DEPA).

In Indonesia, where urbanization and population growth are rapid, the national development plan promotes smart city development in various parts of the country to address challenges such as environmental pollution (Mahesa et al., 2019). In Vietnam, the need for smart cities has been announced by the prime minister and others, although there is no explicitly stated national strategy for the country as a whole. Smart city development is being promoted by the private sector, including foreign-invested enterprises.

In SEA countries, the term "smart city" covers a wide range of scopes. Thailand's DEPA has established seven "Smart Pillars" to categorize the effects on cities, which vary from data-intensive initiatives to infrastructure upgrading and the construction of new residential areas. In developed countries, "smart cities" are often positioned as carbon neutral or low-carbon in their context and significance, while in SEA countries they are often positioned as smart cities for economic growth and industrial promotion rather than the environment.

In any case, the government's passion and interest in making cities and infrastructures smarter makes this a great opportunity to get involved in various reforms. It is important to emphasize here that ICT is only a means or a tool, so it is not just about introducing smartphone apps or electronic payments. The rise of smart cities and the evolution of ICT technology should be used as an opportunity to review the underlying ecosystem to make it more efficient, less burdensome and more convenient for users, policy makers, and transportation operators.

In the next sections, some emerging technologies in the transportation sector, which are also positively affecting smoother and more comfortable mobility in SEA, are discussed. In 7.2, the new technologies and services based on the revolution of automotive last years called CASE are discussed. In 7.3, the raising sharing services which represent "S" in CASE, especially the situation and potential of ride-hailing and ride-sharing, are introduced. In SEA, now grab or other ride-hailing services are more than transportation app but infrastructure for daily life. 7.4 is about Mobility as a Service (MaaS), which is also another revolution of mobility sector in the last ten years. Introducing the concept of MaaS, the integration of each transportation mode is more encouraged and to accelerate the integration, the new ideas on ticketing, booking, and information sharing are also developed.

7.2 CASE

Yurie Toyama

CASE is an acronym for four recent innovations in the automotive industry: C: Connected refers to being connected by communication; A: Automated refers to automated driving; S: Shared refers to the increasing variety of ways in which cars are used, not just owned, such as car sharing and ride hailing; E: Electrification refers to electrification.

Originally presented at the 2016 Paris Motor Show as the Mercedes-Benz Group's medium-term vision, it has become a common noun. With changing values toward cars, for example, the widespread use of cars that are not ownership-oriented, and the changing nature of information and data due to the widespread use of smartphones, car manufacturers are finding that car sales will not grow in the same way as before. They are faced with the need to think about how to adapt their vehicles to the new era and the need to move away from the value of making cars. The CASE is an expression of the need to think comprehensively about how to make the experience of mobility highly satisfying, safe, secure, and smooth, and to be reborn as a mobility service provider.

Furthermore, as the whole world is required to reduce its environmental impact more and more, the transportation industry, which is not a small emitter, has a heavy mission to fulfil and is required to switch to vehicles with a lower environmental impact and to change to a new means of transportation. In this way, mobility is undergoing a major transition, and in parallel, the innovation and rise of various technologies are remarkable. I would like to describe what is happening in technology and mobility now, in line with the key words of each CASE.

7.2.1 Expectations of the Connected

Connected, as the name suggests, is the connection between a car and something else. By connecting the car to the outside world, the Internet of Things can be incorporated into cars. The connectedness of vehicles can be divided into two main categories: safety purposes and convenience purposes.

For safety purposes, for example, it may be possible to support safe driving with Vehicle to Everything (V2X), which literally means the connection between cars and things other than cars. Specifically, this could include cars and each other, cars and infrastructure such as sensors and signals on the road, cars and networks such as the internet, cars and pedestrians, and many others. Although the sensors and cameras installed in cars have continued to evolve and are now able to recognize various situations in the surrounding area, information on obstacles outside the detection range of sensors and cameras, the state of the road on the driving route, and information on signal changes that will occur in the future, for example, can be obtained as V2X from the car's communication partner, which can then be used for safe driving. If the vehicle can receive this information, it will be useful for safe driving. This will also be of great help when looking at automated driving in the future, as on-board sensors and cameras cannot detect all information.

There are several communication methods for V2X: for low-latency communication requirements, vehicle-to-vehicle and vehicle-to-infrastructure can communicate directly using dedicated frequency bands. Another possible method is to utilize mobile lines to connect to the network. The former are currently available technologies, such as IEEE 802.11p (localized in Japan as ITS connect), a type of Wi-Fi, also known as Dedicated Short-Range Communications (DSRC), and cellular V2X, a feature of 5G that has recently become feasible. DSRC has been the subject of various studies and market research, including a series of federal government-led demonstrations in the first half of 2010, the allocation of dedicated frequency bands, and the installation of V2X functionality in some Cadillac models (Kenney, 2011). Various studies and preparations for market launch were carried out, including the installation in Cadillac models. In Japan, 760 MHz has also been allocated for DSRC and is being used in some models of Toyota passenger cars and ambulances in some regions. Recent innovations in 5G technology have made it possible to use direct communication technology alongside 5G mobile lines, including the use of millimeter waves, which may change the trend as to which technology will gain majority market share in the future (Sakaguchi et al., 2021).

For either technology to be truly effective, the more people it can connect with, the better, which is the biggest hurdle for V2X. There are challenges such as cost and space to install V2X-compatible antennas and communication chips in order to penetrate more vehicle models. In addition, for car manufacturers to decide to install V2X in more cars, they would probably consider that it depends on the situation and how many V2X infrastructures are available on the streets. In order to deploy V2X in roadside infrastructure, various policy, budgetary, and technical challenges for road and traffic managers need to be resolved. Thus, it is often pointed out that the diffusion of services involving infrastructure and other vehicles is a "chicken-and-egg problem", so a policy decision is needed where one of them has to be hungry first and start one step ahead in the diffusion process.

Connected for convenience purposes, for example, car manufacturers are proposing connected services that allow users to enjoy music distribution and message exchange from within the car, suggest destinations and other information based on personal characteristics, and request support from an operation centre or other service centre in the event of an emergency. Furthermore, operating systems for linking smartphones with in-vehicle functions (e.g., Apple's CarPlay) are also widely used today.

Another connected technology is the use of communication between the vehicle and the infrastructure for congestion charging, as in Singapore's ERP (Electronic Road Pricing). Due to the high installation rate of in-vehicle equipment, congestion can be monitored with a high degree of accuracy, and charging can be changed according to congestion conditions. In Japan, the widespread use of ETC 2.0 has enabled the system to be used not only for simply collecting tolls on motorways, but also for providing road conditions and measuring traffic volumes. In the future, it is also expected to play a role in assisting automatic vehicles to merge and provide information. Other use cases are also available in the market, such as some health insurance companies and automotive companies offering the price incentive based on the driver's behavior, which is recorded and shared via connected services. Connected technology offers safer and more convenient driving experience.

7.2.2 Expectations for Automated Driving

In developed countries, which are suffering from labor shortages due to declining and ageing populations, they have high expectations for automated driving, particularly of commercial vehicles such as buses, taxis, and trucks, because the number of people required to operate them can be reduced and the burden on drivers can be less than that in the past. However, although the image of automated driving as driverless tends to prevail, benefits can be found even without such a high degree of automation. Rather, advanced driver assistance systems (ADAS) can also be seen as one of the factors leading to automated driving, such as the improvement of traffic flow through the use of Cooperative Adaptive Cruise Control (CACC) and more efficient energy consumption, which can also be beneficial enough in emerging countries (Avedisov et al., 2022).

Examples of self-driving bus demonstrations in SEA countries include the city of New Clark in the Philippines and in the campus of Chulalongkorn University in Thailand (Narupiti et al., 2023). In both cases, the vehicles were driven in areas similar to private property, where driving conditions and the surrounding traffic environment can be controlled to some extent. In terms of automating driving, the evolution of in-vehicle equipment such as sensors and cameras, and the evolution of image processing and AI to link the information detected by these devices to driving behaviour, is advancing day and night. The technological feasibility of automated driving within the scope of the Operational Design Domain (ODD), which defines the technically feasible range of these devices, is increasing considerably.

On the one hand, many other issues remain when considering the application of automated driving to urban transportation, particularly buses and taxis. One of the largest challenges is the need to reduce the price and cost-sharing of automated driving systems. As mentioned earlier, cameras and sensors are evolving, and equipment sufficient to realize automated driving is becoming more achievable, but whether this can be achieved at a price that allows widespread use is another matter. In addition to hardware such as cameras and sensors, it also takes a certain period of time and costs to generate and calibrate digital maps of driving routes. The current situation is very different from the time, amount of technology, and cost it takes for a bus company to purchase a vehicle and be able to start operating it, and the market needs type-approval automated vehicles to secure the safety of the vehicle. Lower prices also lead to the "chicken-egg problem", whereby it is difficult to make progress unless the market expands. The challenge for the spread of self-driving buses and taxis in emerging countries will be to make them even more affordable. Of course, there is hope that prices will come down, but the introduction of automated buses should also be an opportunity to think about how vehicles are purchased and owned, and ultimately about the division of labor between the management and operation of urban transportation services, in a way that moves away from conventional thinking. Traditionally, for example, privately and privately owned transportation operators in SEA countries have been both owners and users of vehicles. However, ownership could be leased, for example, or ownership could be assumed by a regional or wider vehicle management corporation, which would enter into a contract of use with the transportation operator. In Europe and North America, there are a number of private companies that are responsible for public transportation operations under contract from local authorities. Examples include Transdev, Keolis, and First. These companies are taking advantage of the economies of scale of doing business worldwide, and are investing in, and collaborating with, automated driving-related companies and others, in anticipation of automated driving in the future. For example, in situations where the introduction of automated driving will force the initial costs of urban transportation operations to be higher than before, it may be a good idea to consider separating the vehicle retention scheme from the operation and management of the vehicles.

There are examples of automated buses and other vehicles being driven on public roads in Europe, the United States, Japan, and other countries, but for these to be used on public roads, it will be necessary to create an environment that is not straightforward, such as improving road infrastructure, developing legal systems, and fostering social acceptance in the surrounding areas, which will require long-term thinking.

7.2.3 Expectations of Shared

The value of not being concerned with "ownership" as a way of using a car is spreading to a wide range of regions and generations, partly due to the rise of the sharing economy, where people share various things other than cars.

App-based car-dispatch services have spread widely across the world over the past few years, and SEA countries are no exception. In North America, so-called Transportation Network Companies (TNCs), such as Uber and Lyft, started with private carpooling and in recent years have proposed their own multi-modal routing packages and expanded into the sharing of vehicles other than private cars, such as e-scooters and bicycles. This contributes to the provision of multimodal transportation services. It is no longer just an app-based vehicle dispatch service, but is fully functioning as a mobility-related infrastructure.

In SEA, Grab is the biggest player, with local services such as Gojek in Indonesia, and Yandex and Gett in Central Asia and the Middle East. The most popular apps vary from country to country and region, but the general functionality is the same.

What does the introduction of shared service in SEA mean? In SEA countries, there are many paratransits to connect mass transit networks or as an alternate to mass transit networks. However they often face difficulties in catching them on the road or negotiating the fare, which tends to put the service providers in a stronger position than the users. With app-based vehicle dispatch services, it is now possible to be dispatched without having to wait on the road, and fares are determined using logic based on the balance of supply and demand, making it much easier for users to get around. In addition, these car-dispatch apps have made users feel safer than ever before, as they can see how the driver has been driving, based on ratings from other users, and at least know the driver's face photo and nickname. The risk of being involved in various crimes is a concern, especially when women use the service alone, but at least they can now check which driver they have received service from, which is likely to have a significant effect on crime suppression. Furthermore, by becoming part of a larger service platform, drivers are provided with safety equipment such as uniforms and helmets, which has helped to give them a greater awareness of road safety and a sense of belonging and professionalism.

In terms of infrastructure, it is becoming a public digital infrastructure in terms of data utilisation. In Cambodia, an agreement was signed with the Ministry of Public Works and Transport (MPWT) in 2017, when Grab was introduced, to utilize Grab data in terms of road safety and traffic management. Until now, it has been necessary to install infrastructure such as TRACANs or conduct large-scale statistical surveys in order to understand the actual traffic volume and travel demand, but by making use of data from a company with a high penetration rate in the private sector and a high public profile, the government can make use of the data through an agreement, which can also be used for policy This is the phase of the project. Car-sharing or bicycle sharing is also a kind of "sharing". However, authors put more details on ride-hailing and sharing are discussed in 7.3.

Figure 7.1 A Grab driver with helmet and uniform

Source: (Photo taken by Yurie Toyama)

7.2.4 Expectations for Electrification

The electrification of vehicles has been rapidly promoted worldwide in recent years to reduce environmental impact. This rapid increase in demand for electric vehicles (EVs) has been positioned by many SEA countries in their policies as a catalyst for industrial development and economic growth. On the one hand, leaving aside the wealthy who are able to own electric passenger cars as private vehicles, the electrification of public transportation such as buses and taxis poses many challenges for transportation operators in the field, such as the high initial cost burden and hurdles in the development of associated infrastructure such as charging facilities.

Thailand and Indonesia are probably the two most representative countries that are focusing on electric vehicles as an industry. Thailand has always had many factories related to the automobile industry, but in order to focus on EV production, it is attracting industry by offering tax incentives to EV and related component manufacturers. Indonesia has the world's largest reserves of nickel, the raw material for EV batteries, and aims to utilize this resource to develop its EV industry. All of these countries have targets, such as the EV share of new car sales in the country, but in reality the price of

EVs is still effective, and the range of people and companies that can afford them are limited.

Nevertheless, there is an urgent need to reduce GHG emissions from the transportation sector, especially in large cities with high population concentrations and heavy traffic, and high expectations for electrification.

For example, Transjakarta, the BRT system that provides backbone transportation for Indonesia's largest city, Jakarta, will start introducing electric vehicles in 2020 and has a target to electrify all of Transjakarta's over 10,000 vehicles by 2030.

Jeepneys, an important means of transportation for Filipino residents, are also increasingly being electrified. The government has urged a switch to electric or less environmentally hazardous vehicles, with the aim of reducing GHG emissions from jeepneys. However, many jeepney operators are individuals or family-owned businesspersons and cannot easily replace their vehicles. The switch to electrification is not sustainable unless it is considered from many angles, including a purchase support system to make it easier for operators to purchase vehicles, sharing new engineering knowledge as they go electric, developing infrastructure such as charging facilities, and developing an operation plan based on charging.

7.2.5 Potential of CASE in SEA and Emerging Countries

In Section 7.2, we discussed services and use cases that make use of each of these technologies, with a view to their application in emerging countries, in line with the CASE (C: Connected, A: Automated, S: Shared, E: Electric) keywords that describe the major changes that have taken place in the automotive industry in recent years. "Connected" has the effect of enabling things that were previously impossible with a single car alone, by connecting cars to the internet, to other cars, and to roads. For example, it is possible to receive information that cannot be captured by in-vehicle sensors or cameras from outside the vehicle, and by constantly receiving the latest information via the internet, it is possible to reduce the risk of accidents by preventing many accidents, and to enjoy entertainment and drive comfortably. "Automated" is the automation of driving. Automated driving is expected to provide solutions for securing a sufficient number of professional drivers by reducing the burden of driving, and to improve safety by reducing the burden of relying on human perception and judgement. "Shared" reflects the growing need for car use that is not focused on ownership in an era when the sharing economy is expanding beyond cars. "Electric" cars are desired from the perspective of the global urgency of addressing low-carbon issues and their compatibility with new technologies such as automated driving and connectivity. In emerging countries too, there are many cities where reducing traffic accident damage and eliminating traffic congestion by improving car safety are an issue. To solve these issues, it will contribute not only to improving the performance of the car itself, but also to helping people to own, use, and drive cars more intelligently and comfortably.

7.3 RIDE-HAILING

Mihoko Matsuyuki

7.3.1 Introduction

As mentioned in 7.2, ride-hailing (RH) is currently experiencing explosive growth in the number of its users in the emerging Asian countries, with an estimated 1,139 million users in the region by 2024 (Statista, 2024a). Especially in SEA, the number of RH users is growing rapidly, with 40 million active users in 2019, and its revenue is estimated to reach $8.87 billion by 2024. Its compound annual growth rate (CAGR, 2024–2029) is 5.39%, and the market size is projected to reach US$ 11.53 billion in 2029 (Statista, 2024b). In SEA, Grab and Gojek, two of the 11 RH unicorn companies in the world, dominate the market; Grab and Gojek are of Singapore and Indonesian origin, respectively, and are decacorn companies with market values of US$ 14.9 billion and US$ 12.5 billion. Grab has 183 million users and 2.8 million drivers, while Gojek claims to have 170 million users (app downloads) and 1 million drivers (Chalermpong et al., 2022). In addition, many other local RH companies operate in SEA cities. Thus, RH is now rapidly increasing its users in Asia, especially in cities in emerging Asian countries, and has the potential to change urban transportation systems. In addition, different types of RH are operated in Asia than in North America, where RH was originally developed, and the way in which RH is used in Asia is different from that in North America. Therefore, this section describes the current status, challenges, and future development of RH in emerging Asian countries, especially in SEA, where the use of RH is flourishing.

7.3.2 Characteristics of RH in SEA

7.3.2.1 RH Platform in SEA

Uber first started its RH business in 2013 in SEA, and shortly after, Grab began dispatching four-wheeled RH vehicles. Gojek then began dispatching motorcycle RH in 2015; Uber sold its business in SEA to Grab in 2018 and exited the region. Thus, in SEA, local companies have dominated the RH market, led by two locally born decacorn companies. Now Grab operates its RH business in eight countries and Gojek operates in four countries in SEA. One of the reasons for this is that Grab and Gojek, in particular, offer a variety of services on their apps, including online food delivery, online grocery shopping, e-money, and courier services, in addition to RH. Food delivery is one of the most widely used of these services.

7.3.2.2 Paratransit RH

While the majority of RH in North American countries are by car-type, a variety of paratransit RH can be seen in SEA. For example, Grab offers motorcycle RH in Indonesia, Vietnam, Thailand, and tuktuk (three-wheeled cab) and

Figure 7.2 Tuktuk type electronic ride-hailing in Bangkok

Source: (Photo taken by Mihoko Matsuyuki)

remorque (motorcycle towing a passenger car) RH in Cambodia, while other local RH companies also offer paratransit RH services (Figure 7.2). Among these paratransit RH services, motorcycle RH is used more than four-wheeled RH in most SEA countries, especially in countries with a large motorcycle share, such as Vietnam, Indonesia, and the Philippines. Motorcycle RHs are used more because their small size allows them to slip through traffic jams and they are cheaper than four-wheel RH.

As discussed in Chapter 4, paratransit in emerging countries has various issues such as poor vehicle maintenance, poor quality of service, and safety issues. When paratransit is combined with RH, it is expected that some of the RH's issues will be addressed. For example, RH has an incentive mechanism where passengers evaluate drivers and their evaluations affect the receipt of future dispatch requests and driver rankings, which in turn affect their income and benefits provided by RH operators. This incentive mechanism motivates drivers to maintain their vehicles and to further improve the quality of their service, including safe driving and polite responses to passengers. In addition, paratransit is flexible, but there is uncertainty of arrival time and price is decided by negotiation in some types of paratransit. The integration of paratransit into RH will allow users to check the status of their dispatch via an app, and the price will be clearly indicated when they request a ride. Also, while conventional motorcycle taxi passengers usually ride without helmets, both Grab and Gojek equip a helmet for a passenger and require not only drivers but also passengers to wear helmets, and both drivers and passengers are covered by insurance in the event of an accident.

In this way, some of the issues that paratransit faces have been addressed by incorporating it into RH. The combination of paratransit with RH has increased convenience and improved the service and safety, resulting in a large number of users.

7.3.2.3 Complementation and Substitution

One of the controversial issues regarding RH is which mode of transportation it substitutes or complements. Compared to four-wheel RH, paratransit RH is characterized by its shorter mileage. While four-wheel vehicles in North America often substitute taxies, the paratransit type RH often substitutes walking or public transportation (Paronda et al., 2017). In addition, a high percentage of paratransit types are also used as first- and last-miles of public transportation, complementing medium- and mass transit (Chalermpong and Ratanawaraha, 2015; Irawan et al., 2019).

It has also been reported that motorcycle ride-hailing is used more by women (Silalahi et al., 2017), providing mobility to people, who do not have driving licenses, cars, or motorcycles, which has the effect of improving equity in terms of mobility.

7.3.3 Challenges of Ride-Hailing in SEA

As mentioned earlier, the rapid spread of RH has had a positive impact on the transportation system, as it has greatly increased convenience for people and improved accessibility to public transportation. However, there are also many challenges associated with RH.

First of all, RH has been criticized for exacerbating traffic congestion by increasing traffic volume as it replaces walking and public transportation (Icasiano and Taeihagh, 2021). Furthermore, paratransit RH is often used as a first/last-mile alternative to medium- and mass transit systems such as rail and BRT. Therefore, during peak hours such as morning and evening, many RHs gather around stations, exacerbating traffic congestion. Furthermore, environmental issues such as air pollution and GHG emissions caused by the increased traffic and traffic congestion created by RH have also become an issue.

In addition, the working conditions and working environment of RH drivers, as well as the risk of accidents and conflicts with existing transportation operators such as taxies and the conventional paratransit system, are also raised as issues (Radzlan et al., 2023; Nussara et al., 2024).

Thus, the rapid diffusion of RH has many positive and negative impacts, and RH needs to be properly regulated and utilized in order to establish a sustainable transport system. Nevertheless, only a limited number of countries have a regulatory regime for RH, and even in those countries, the regulation of RH is still in flux (Chalermpong et al., 2022).

7.3.4 Future of Ride-Hailing in Emerging Asian Countries

One of the future directions for RH in Asia is integration into MaaS. Already the Grab and Gojek apps have a trip planner function, which is at the level 1 stage of MaaS. Furthermore, JakLingko, which is the current MaaS app for Jakarta, includes Grab in addition to trains and BRT, and allows the purchase of tickets, which is at the level 2 stage of MaaS. Thus, the RH app itself is becoming a MaaS platform due to its multi-functionality, and RH is also being incorporated into the MaaS system.

Next, we need to consider how to regulate RH and how to incorporate RH into government's transportation plans. As mentioned earlier, the regulation of RH is in flux; we need to identify the impact of RH on the transportation system, the environment, and so on, and regulate it appropriately to make the city's transportation system sustainable. And since RH has become an indispensable travel mode in SEA and is used as the first/last mile mode of medium- and mass transit, it has the potential to make a significant contribution to the development of a sustainable transportation system. Therefore, RH should be positioned in transportation planning and should be used to promote the use of public transportation.

7.4 MAAS

Yurie Toyama

Mobility as a Service: "MaaS" is a term that became popular from Europe around 2014. In recent years, the term "mobility" has been used in some cases as "vehicle", which is not very welcome in terms of trivializing the scope of what mobility means. Essentially, mobility means ease of movement and freedom of movement. MaaS can then be understood as the provision of mobility as a service. If understood in this way, MaaS is not just a smartphone application, but a multi-modal concept that provides a one-stop shop for all information and payment methods, transcending the boundaries between transportation modes and operators.

To begin with, the high penetration of smartphones in South-East Asia has a high affinity for the introduction of MaaS. Various services are being deployed day and night in all cities around the world, but here we organize our perspectives on how to utilize MaaS as urban transportation planners and policy makers.

It was noted earlier that MaaS is more than just an app. However, by using ICT technology, bundling various transportation services, and often being described as an idea linked to industries other than transportation, MaaS has undoubtedly provided an opportunity for parties in the urban transportation sector to talk to a wider variety of people, and this movement can and should be capitalized on. Furthermore, it could also provide an opportunity for developed countries to review their traditional operations and redesign passenger services in a more digital data-orientated way, in light of the labor shortages in the transportation industry, for example.

7.4.1 MaaS as a Demand Management Tool

MaaS can be a demand management tool, depending on how information is given to citizens and how pricing is set. For example, there are situations where governments and transportation operators do not want users to use certain transportation systems, sections, or times of the day because of congestion, maintenance, weather conditions, and so on, or because they do not want to concentrate users. At such times, demand can be dispersed by, for example, providing information on alternatives to avoid congestion and trouble, or offering lower fares if a diversion is made. In addition to these temporary measures, it is also possible to encourage people to take action by, for example, promoting restrictions on private cars in city centres through park-and-ride, or by adding more price-attractive menus to encourage carpooling, by devising ticketing according to recommended transportation mode combinations and conditions of use.

7.4.2 Improving the Efficiency of Passenger Services

MaaS can also be seen as a catalyst for so-called digital transformation (DX), such as the digitalization of analogue data, the digitalization of data and systems that were independent and not linked, and the provision of data to passengers that could not be provided. For example, real-time information on the position of operations and information on delays can be provided to passengers. Transportation operators can make it as much as possible a one-stop shop to share information on timetable changes, decisions on cancellations, and delays within the operator with station staff and conductors, and to share this information with users. If this information is completed on a smartphone, it could, for example, make it easier to respond to users at the station counter or over the phone.

7.4.3 Improving High Added Value through Cooperation with Other Sectors

Japanese MaaS, which is evolving in its own way, includes ideas such as tourism MaaS, in which transportation operators are linked to tourist attractions, and medical MaaS, in which transportation operators are linked to medical facilities. The main source of revenue for transportation operators to date has been fare income, but services combining excursion plans with tickets, for example by proposing excursion destinations, offering coupons for excursions, and advertising income for this, can add value by providing a better experience from mere travel. This may also improve the added value. In emerging countries, this could involve, for example, frequently changing the fare structure for residents and foreign tourists, or offering higher value-added products for foreigners.

7.4.4 Expectations of MaaS in Emerging Countries

In Section 7.4, we discussed the potential of MaaS. In many SEA countries (such as Thailand, the Philippines, Indonesia, and Malaysia), there is a mix of different operators, including national railway operators that connect inter-city railways, operators that run subways and buses within cities (it is not uncommon for different companies to operate different lines or modes within a single city), and paratransit operators, and in many cases, information sharing and ticketing systems are independent. However, increasing the coordination between modes and integrating them as a single service will increase the value for users in terms of convenience. Furthermore, the value of MaaS will not only be for users, but also be an opportunity for operators to digitize various data so that it is compatible with digital services, to change their operations to be based on the use of digital data, and to improve their services based on data.

7.5 CONCLUSION

Yurie Toyama

Chapter 7 discusses new technologies, service models, and concepts that have brought transformation to the mobility sector due to advancements in ICTs in recent years. The chapter examines their potential applications and effects in emerging countries.

In Section 7.1, the widespread use of smartphones is highlighted as a background factor enabling the diverse and accessible application of ICTs. It points out that in SEA, the so-called leapfrogging phenomenon led to the proliferation of mobile phones, surpassing fixed-line telephones, resulting in a high mobile phone ownership rate. High individual ownership of mobile phones facilitates the provision of ICT-enabled services, as well as data collection and utilization, making it easier to meet the mobility needs of individuals. Additionally, amid global trends such as Society 4.0 and decarbonization, governments across SEA are pursuing initiatives related to smart cities to create smarter, more energy-efficient, and low-emission nations and cities. As various administrative services and policies undergo a digital shift, the transportation sector is also transitioning to a more digital and data-friendly approach that caters to specific and detailed needs. This shift not only enhances the efficiency, value-added services, and decarbonization of transportation businesses but also fosters smoother integration with various urban activities.

In Section 7.2, the discussion revolves around the CASE framework – Connected, Autonomous, Shared, and Electric – highlighting respective technologies and their applications or use cases in the automotive industry. While these might initially seem like topics solely relevant to high-end private cars, this is not the case. Even technologies like Level 2 autonomous driving (driver assistance) just below Level 4 full autonomy can aid human recognition and decision-making, assisting in optimal spacing and speed, and delivering tangible benefits. CASE technologies have the potential to bring significant innovations not only for vehicle owners and drivers but also for passengers utilizing vehicles. In emerging countries, where traffic accidents and congestion are major urban issues, improving vehicle safety and resolving these problems are critical goals. These advancements not only enhance vehicle performance but also contribute to smarter, more comfortable ownership, utilization, and driving experiences.

Section 7.3 focuses on RH and ride-sharing services, which are part of the "S" in CASE. RH in SEA has seen significant usability improvements, moving away from street-side hailing and price negotiation to app-based systems where users can book rides and pay fixed prices. The introduction of app-based platforms has also enabled driver rating systems, improved labor management efficiency, and opened up new employment opportunities for drivers. In SEA, RH app systems have extended their functionality to include food delivery and QR code payments, becoming an integral part of daily life infrastructure beyond mere transportation.

In Section 7.4, the potential of MaaS is explored. Many SEA countries feature diverse operators for intercity railways, urban subways, buses, and paratransit systems, often with independent ticketing systems and unshared information. Enhancing integration among modes and unifying them into a single service-like platform increase convenience and value for users. Historically, challenges like limited access to information and complex fare systems posed hurdles for non-residents in utilizing public transportation. However, the introduction of MaaS and the sharing of operational data with international services like Google Maps have lowered these barriers, making public transit more accessible to tourists and captive users of public transportation.

In emerging countries, it is common to receive support in the form of technical assistance or goods from developed countries. However, merely introducing new technologies in isolated areas has limited effects on broader trips and lifestyles. As discussed in this chapter, it is crucial to consider the various characteristics, use cases, and intended impacts of new technologies from both the passenger's and operator's perspectives. New technologies should be tools to be used effectively, not ends in themselves, and they should not disrupt human lives.

REFERENCES

Avedisov, S. S., Bansal, G., & Orosz, G. (2022). Impacts of connected automated vehicles on freeway traffic patterns at different penetration levels. *IEEE Transactions on Intelligent Transportation Systems*, *23*(5), 4305–4318. doi:10.1109/TITS.2020.3043323

Chalermpong, S., Kato, H., Thaithatkul, P., Ratanawaraha, A., Fillone, A., Hoang-Tung, N., & Jittrapirom, P. (2022). Ride-hailing applications in Southeast Asia: A literature review. *International Journal of Sustainable Transportation*, *17*(3), 298–318. doi:10.1080/15568318.2022.2032885

Chalermpong, S., & Ratanawaraha, A. (2015). How land use affects station access behaviors of bus rapid transit passengers in Bangkok, Thailand. *Transportation Research Record: Journal of the Transportation Research Board*, *2533*(1), 50–59. doi:10.3141/2533-06

Icasiano, C. D. A., & Taeihagh, A. (2021). Governance of the risks of ridesharing in Southeast Asia: An in-depth analysis. *Sustainability*, *13*(11), 6474. doi:10.3390/su13116474

Irawan, M. Z., Belgiawan, P. F., Tarigan, A. K. M., & Wijanarko, F. (2019). To compete or not compete: Exploring the relationships between motorcycle-based ride-sourcing, motorcycle taxis, and public transport in the Jakarta metropolitan area. *Transportation*, *47*(5), 2367–2389. doi:10.1007/s11116-019-10019-5

Irvine, K. N., Suwanarit, A., Likitswat, F., Srilertchaipanij, H., Ingegno, M., Kaewlai, P., Boonkam, P., Tontisirin, N., Sahavacharin, A., Wongwatcharapaiboon, J., & Janpathompong, S. (2022). Smart City Thailand: Visioning and design to enhance sustainability, resiliency, and community wellbeing. *Urban Science*, *6*(1), 7. doi:10.3390/urbansci6010007

Kenney, J. B. (2011). Dedicated short-range communications (DSRC) standards in the United States. *Proceedings of the IEEE*, *99*(7), 1162–1182. doi:10.1109/JPROC.2011.2132790

Mahesa, R., Yudoko, G., & Anggoro, Y. (2019). Dataset on sustainable smart city development in Indonesia. *Data in Brief*, *25*, 104098. doi:10.1016/j.dib.2019.104098

Narupiti, S., Phetcharavut, Y., & Pakpisutkul, R. (2023). Perception of prospective users on autonomous shuttle as a smart mobility service: A case of Chulalongkorn University, Thailand. In *2023 IEEE PES 15th Asia-Pacific power and energy engineering conference (APPEEC)* (pp. 1–6). IEEE. doi:10.1109/APPEEC57400.2023.10561983

Nussara, T., Joseph, B., Tim, A., & Kurtis, H. (2024). "I will just have to keep driving": A mixed-methods investigation of lack of agency within the Thai motorcycle rideshare driver community. *Proceedings of the ACM on Human-Computer Interaction*, *8*(CSCW1), 1–28. doi:10.1145/3653706

Paronda, A. G. A., Regidor, J. R. F., & Napalang, M. S. G. (2017). Service characteristics of transportation network companies (TNCs) and conventional taxis in Metro Manila, Philippines. *Journal of the Eastern Asia Society for Transportation Studies*, *12*, 2457–2472.

Radzlan, R., Othman, M. A. N., & Yusoff, I. M. (2023). The impact of working environment on health problems among e-hailing drivers in Johor Bahru, Malaysia from ergonomic perspective during pandemic COVID-19. *International Journal of Academic Research in Business and Social Sciences*, *13*(18), 1–15. doi:10.6007/ijarbss/v13-i18/19966

Sakaguchi, K., Fukatsu, R., Yu, T., Fukuda, E., Mahler, K., Heath, R., Fujii, T., Takahashi, K., Khoryaev, A., Nagata, S., & Shimizu, T. (2021). Towards mmWave V2X in 5G and beyond to support automated driving. *IEICE Transactions on Communications*, *E104.B*(6), 587–603. doi:10.1587/transcom.2020EBI0001

Silalahi, S. L. B., Handayani, P. W., & Munajat, Q. (2017). Service quality analysis for online transportation services: Case study of GO-JEK. *Procedia Computer Science*, *124*, 487–495. doi:10.1016/j.procs.2017.12.181

Statista. (2024a). Number of users in the ride-hailing and taxi market in Asia. *Statista*. Retrieved from https://www.statista.com/forecasts/1124692/number-of-users-in-the-ride-hailing-and-taxi-market-in-asia

Statista. (2024b). Ride-hailing market in Southeast Asia. *Statista*. Retrieved from https://www.statista.com/outlook/mmo/shared-mobility/ride-hailing/southeast-asia

Chapter 8

Modernization of Passenger Transportation Management

8.1 INFORMATION PROVISION INTEGRATION

Toshiyuki Okamura

8.1.1 Information for Transit Users

The ability of users to know where to board, the route, the transportation mode, and the transfer point is fundamental to the ease of travel in passenger transportation. It is important that the necessary and accurate information is provided to users, that it is easy for them to access, and that the information is easy for them to understand. Furthermore, it is desirable that the information provided corresponds to the various usage scenarios of various segments of users. Especially in the case of public transportation, appropriate and easy-to-understand information provision will not only increase convenience for frequent users of public transportation, but also increase the attractiveness of public transportation for potential users, such as visitors and private transportation users, leading to an increase in the number of public transportation users and a reduction in private transportation use.

The basic information needed by intra-city travelers is static information, such as the place of embarkation and disembarkation (stations, stops), fares and payment methods, timetables, vehicle destinations and transit stops, routes, transfer points, and maps (route maps and layouts around destinations and public transportation interchanges) for public transportation.

Dynamic information such as actual or estimated travel times, congestion levels, and real-time vehicle locations, as well as irregular operations such as suspensions, will be added. In addition to mass transit, cabs and ride-hailing availabilities, private cars, motorcycles, and bicycles will also require information on routes, congestion, and parking locations. People are using a variety of information to select and compare the best means and routes for getting around the city, sometimes combining a complex range of transportation modes. The ability of people to successfully use urban transportation depends largely on the quality and quantity of information

DOI: 10.1201/9781003467403-10

provided, the means of information provision and its availability, and information literacy on the part of users. The key to the modernization of the first two is the coordination and integration of information provision and digitalization.

8.1.2 Acquisition of Transit Data

In the case of public transportation, most of the information provided is related to transportation supply, such as timetables, routes, and real-time locations. In other words, the information here is about the transportation service providers. The first challenge in emerging countries is how to develop this information, that is, basic static information, into visualization and digitization. Informal public transportation, such as paratransit, is composed of countless small private entrepreneurs, who are vehicle owners and their associations, and cannot be called "transportation operators". Even "bus operators" often operate on a commission basis with drivers and fare collectors, and the concept of operational management itself may be small or almost non-existent. Before the integration of information, the transportation provider does not have a grasp of operational information. In cases where there is a registration or licensing system by the government, the government is aware of the registration information of routes, vehicles, drivers, and so on of private transportation providers and their associations, but in many cases, the registration contents are not digitized or do not reflect the actual situation.

In other cases, the public transportation service itself may be in its infancy, and the information itself may not be available (e.g., the boarding point and operating hours are not fixed, and the frequency of the service depends on the number of users). To begin with, there is a need to formalize the operations and transportation services of informal transportation providers and private operators, as well as to clarify and digitize the notification, registration, and licensing information to the government in the process.

Second is how to obtain more advanced information. For dynamic data (e.g., congestion information and real-time location), it can be said that these information are rather easier to acquire than static information, as GPS equipment and communication costs have dropped significantly. Advanced information is also the newly arisen static information (e.g., detailed boarding location information, timetable information, detailed route maps, information about transfers) as public transportation services become more formalized and sophisticated. Third is who will coordinate and integrate public transportation information. These are information that formalized operators maintain themselves and must maintain on their own.

Third is who will coordinate and integrate public transportation information. Public transportation information is being developed as open data, and a common data format for public transportation called the General Transit Feed Specification (GTFS) is becoming popular worldwide. The GTFS defines formats for static and dynamic information, and is used for data utilization and display in a variety of applications. As transportation operators develop and publish their own data in this format, data from multiple operators is integrated and used in applications such as maps and route finders on the Internet to provide users with integrated information. They are now often provided by private third parties. The integration of urban public transportation operations, especially with the establishment of various transportation associations/transit alliances in European cities since the 1970s and the implementation of BRT mainly in Latin American cities since the 1990s, has been a major factor in the development of urban public transportation systems. The integration of operations has been one of the main driving forces behind the integration and modernization of public transportation services, including the provision of information to users. In both cases, integrated public transportation organizations have taken the lead and focused on integrating and upgrading the information provided to users and improving usability and visibility (TraffiQ and GIZ, 2018). In recent years, the integration of information has expanded to third parties in the private sector, along with the generalization of information technology and the downsizing of information equipment. This has led to a clear trend toward the coordination and integration of information provision, which in turn is driving the transparency, operational efficiency, and modernization of relatively uncoordinated, less formal public transportation services, especially in emerging countries. This is part of a series of moves toward MaaS, as well as the separation of the transportation service provider from the transportation information provider.

8.1.3 Provision of Digital Information and Printed Signages

The fourth challenge in emerging countries is who provides what information and through what channels. This section discusses digital information. If data on public transportation in a city is maintained as open data, for example, map applications such as Google maps or route finder APPs by third parties can be used to obtain integrated information such as time, route, and real-time location information depending on the city and route. In many cases, operators entrust third parties to provide services such as route finder. In other cases, operators, their alliances, or governmental transportation departments often provide their own APPs that provide integrated information on their own. In such cases, the APP provides real-time location information and route search displays that are easier to use and view, and it also handles marketing and fare payment functions. The following are examples as of 2024. City bus authorities in Bangkok, Hanoi, and Ho Chi Minh City provide bus route finder and real-time location information in their APPs. These are more focused on providing dynamic location information rather than static timetable information. Considering the congestion on the roads, dynamic information is more useful to users. In Jakarta, integrated static and dynamic information for MRT, LRT, sub-urban railroads, BRT, and its feeder minibus (microtrans: formerly angkot, an informal paratransit) is provided by JakLingko, a company established by these operators, in its app. JakLingko also provides fare payment by IC card, and has also integrated the fare system, setting a maximum local fare and making microtrans fares free by IC card use (Arif et al., 2023). Manila is still in the process of developing open data, and Google maps, for example, does not support search in many modes and routes. However, a private application provider called Sakay collects and updates data from government agencies and operators, and provides route finder APP for urban rail, bus, and paratransit (jeepney and tricycle) in the metropolitan area. Although real-time location information is limited, the ability to search for routes and fares for the dense and complex jeepney network was a breakthrough.

Finally, what should traditional signage and maps become in the age of digitalization? With many people owning smartphones, even in emerging countries, digital information and the provision of information through digital devices are extremely versatile and applicable. Even if a bus stop has only one pole and no route number or timetable, or even if there is no pole, people can easily use a bus from that location if they can obtain information from an APP on their smartphones and if that information is correct. On the contrary, public transportation systems that have been developed since the pre-DX era have stations and stops that display the names of places and route numbers in iconic logos and fonts, sophisticated and systematic signage based on a unified design concept, and so on, so that the information you want is clearly displayed where you want it. Various innovations have been made to the signage and its location to ensure that the information you want is displayed in a way that is easy to understand where you want it. In addition, easy-to-understand network maps, maps of station neighborhoods, and maps of transportation hubs are also displayed. While their main purpose is to provide information, they also strongly emphasize to people the high level of sophistication of public transportation and the fact that public transportation itself is an icon of the city. The combination of iconic digital and printed signage in the information provision is no less important for the modernization of urban public transportation in emerging countries.

8.2 NETWORK IMPROVEMENT AND INTERCHANGE MANAGEMENT

Fumihiko Nakamura

8.2.1 Importance of Master Plan

As the size of a city expands, the network function of public transportation routes becomes more important. In emerging countries that are large cities but have not yet developed sufficient rail lines, there are conventional rail lines, new rail lines, and bus lines, and the network as a combination of these will be diverse.

Since it is difficult to understand the discussion if the entire network is considered as a whole, this section organizes it into the railroad network, the bus network, and the trunk line branch line network. The railroad network, on the one hand, is a network of railroad lines that are spread over an urban area, and the bus network is a network of bus lines that are spread over an urban area. In most metropolitan areas, the rail network covers a large area but does not have many routes. The bus network, on the other hand, is a large-scale network that covers the entire urban area. The trunk line and branch line network is a network concept in which each of the aforementioned rail and bus networks is considered as a trunk line, and branch lines are established from the main stations of these networks into the district.

Needless to say, this network concept is very important when promoting public transportation planning in large cities. Especially in the case of track systems, whether they are elevated railroads, subways, monorails, or medium-volume track transit systems, they tend to be individual projects for each line because they are large investments in terms of both money and area, the people involved are diverse, and it takes time for planning through completion. Each line tends to have its own independent plan, and construction tends to proceed according to that plan. This is why it is important to have a master plan that summarizes the entire network from a bird's eye view at an early stage. With a master plan, the mutual positioning of plans for individual lines and the division of roles are defined, so there is less blurring and less chance of inconsistency in dealing with uncertainties in individual plans. Even when a later project is proposed, the master plan is the first mover and will have the later project adjusted to be consistent with the contents of the master plan.

If there were no master plan, construction would proceed on an individual project basis, and although the route network would grow, there would be no network effect at all. Compared to railroads, buses require a smaller investment scale and a shorter period of time from planning to operation, making it easier to change or revise routes. Therefore, it is relatively easy to adjust plans for new bus routes. Nevertheless, when considering various facilities including transit facilities and coordination with urban functions as described in the following section, it is better to have a master plan, even if the time span is slightly shorter.

In preparing the master plan, demand forecasting for the target year is required. Calculations will be based on demand forecasting methods developed in developed countries, but in the case of emerging countries, along with the lack of data to begin with, the consideration of uncertainties such as rapid changes in political and economic conditions will be a challenge. Therefore, several points are important to note when forecasting demand for master planning in large cities in emerging countries. First, long-term planning, such as 20 years from now, should be avoided, and the conditions should be simplified as much as possible. Second, it is important to be able to produce forecast results with a wide range of variability, taking into account various uncertainties. This procedure is to use the forecast results to verify the validity of the proposed network, rather than to design a network based on the demand forecast results. Specifically, multiple network proposals are identified based on geographical conditions and the target image of urban development, and demand forecasts are then used to compare these proposals. If a network proposal is predicted to have a significant impact on demand processing during congestion periods, the network is modified, the future layout of urban functions is modified, or the proposal is rejected. It is thus very important to have a master plan that validates quantitative data.

However, as has been said for a long time, in the case of a lattice network, there will be a small percentage of transfers between any two points that require more than one transfer (e.g., a transfer between two parallel lines cannot be achieved with only one transfer). Therefore, it is expected that the shape of the network be designed in such a way that the number of transfers will not be excessive based on the geographical conditions of the city and the layout of urban functions.

8.2.2 Coordination of Service Operation

It is not unusual for different operators to operate different routes, not only in emerging countries but also in developed countries. How to coordinate such services is an important issue. From the user's point of view, coordination of routes, stations, and stop locations, ease of understanding, physical barriers when transferring between different operators, barriers in terms of fares, and barriers in terms of operation schedules (inconsistencies in timetables, lack of consideration when connecting, etc.) should be reduced or avoided. From the operator's point of view, cost savings can be expected through the sharing of facilities and equipment parts. In terms of the policy framework, the above should be synthesized, but it is important that the policy be a viable option for motorcycles and private cars, and that it be financially sustainable.

Services on the public transportation network in urban areas are essentially of high frequency, and barriers in terms of operational scheduling are small. Coordination of route locations and station or stop locations should be completed to some extent during master planning whenever possible. In the case of bus routes, there may be on-site coordination, as routes can be established or changed in the short term. It is desirable that coordination among operators should not be left to the operators, but should be coordinated under the leadership of the government agency in its capacity as the administrator of transportation policy.

In addition to these, particularly serious barriers would be physical barriers and fare barriers. Regarding physical barriers, it is desirable to take the initiative in the master planning stage to make adjustments, and to determine basic policies for facility design and cost sharing. Regarding fares, issues such as common payment media and common fare calculation can be raised. These have evolved greatly, at least in terms of technology, with recent technological innovations, and we can expect various ideas to be implemented even in large cities in emerging countries. The commonization of payment media will progress dramatically with the spread of credit card touch payment. As for the commonization of fare calculation, the digitalization of payment media will greatly improve the efficiency of settlement operations.

8.2.3 Connectivity and Interchange

Needless to say, the effectiveness of a network depends greatly on the function of its nodes. Network connectivity, or connectivity, varies greatly depending on the location on the network, the shape, and the function of the transfer facility, or interchange facility.

Fortunately, in recent years, there are more than a few examples of well-designed transportation nodes, mainly in developed countries, and it is desirable to design them with sufficient reference to these examples. There are also many examples of excellent designs, mainly in Europe, for information displays, which have a significant impact on ease of understanding. These examples can be found in considerable detail on the Internet, so it is advisable to refer to them. It is possible to learn from advanced examples, both in terms of consideration for so-called handicapped people, including wheelchair users, and in terms of universal design.

However, there are several issues that are not easily resolved by simply referring to the case studies and are listed here.

The first is its relationship to the city. Historically, transportation nodes are often located in places with high concentration within a city. In cases where public transportation is heavily used, new transportation nodes may well be expected to attract urban functions and activities. In any case, transportation nodes are not confined to spatial design for transportation functions alone. In their adjacent areas, a variety of urban functions are concentrated and a variety of urban activities are carried out. For example, in many cities in emerging countries, most old intercity bus terminals have so-called markets inside or in adjacent locations. Although it may appear to be a function for killing time in transit, there is no doubt that the hub of people's wide-area traffic is connected to the formation of demand for the market. It is important to read the urban functions.

In Curitiba, Paraná, Brazil, cited several times in this publication, all 10 ward office facilities in the city are attached to transit terminals of the integrated bus route network; in the example shown in Figure 8.1, a bus terminal was also constructed as part of the ward office facility development.

Second is the relationship with roads. Since urban railroads in emerging countries are often built on or under existing trunk roads, traffic nodes are naturally located near trunk roads. As a result, strategic design is required to determine how to connect or not connect traffic nodes to arterial roads. The location of roundabouts, stops, entrances, and exits for buses, cabs, and private cars at traffic nodes is a bottleneck in relation to arterial roads.

Figure 8.1 City Hall branch and bus terminal in Curitiba

Source: (Photo taken by Fumihiko Namakamura)

In addition to this, there is concern that the design will become even more complicated if the movement of traffic used for so-called paratransit is taken into consideration. In many cases, the design of vehicle flow lines for motorcycles and three-wheeled vehicles is not covered in the design documents of developed countries, so it will be necessary to develop an original design based on the actual conditions in urban areas in emerging countries.

Third is the issue of cost sharing. There are several ways of cost sharing. The first method is to share the costs among the public transportation operators directly involved. This is a very natural approach in cities where public transportation has historically been operated under the leadership of private operators. However, if the cost-bearing capacity of the operator is small, the resulting transportation node itself may be very poor, so care must be taken. Next is the use of road revenue sources. This is the concept of designing road facilities as a part of road facilities or designing traffic nodes based on the logic that the design of traffic nodes will lead to the improvement of overall road traffic, utilizing so-called road revenue sources. In some cases, such as Japan's urban monorails and other transportation nodes, this concept is used to leverage gasoline tax revenues. Another way is to utilize the financial resources of the city side. By linking the development projects of adjacent public or private facilities, the city can use the financial resources of those facilities to pay for part or all of the transportation nodal points. The last approach is to have the government pay for the entire cost of the urban facilities. Although this is the most reasonable approach in terms of public transportation in the city, it is not a realistic approach at this time, considering the financial situation of the government.

8.2.4 Mobility Hub

The term mobility hub has been around since about 2015. Specifically, a mobility hub is a small-scale transportation node with car-sharing and bike-sharing stations, electric vehicle charging facilities, and in some cases, delivery boxes, as well as cafes and other spaces for chatting.

In cities in emerging countries, food and beverage stalls spontaneously gather around bus stops that are in demand to some extent, and chairs and tables are laid out on and off the streets, and in some places, motorcycle cab stands are provided. This spontaneous accumulation of functions can be called a substantial mobility hub. An example in Bangkok's Ban Sue district is shown in Figure 8.2.

However, in more than a few cases in emerging countries, illegal use of roads or private land off roads seems to be the case, so it is desirable to take measures for space management and, if possible, to secure public space where such use is allowed in the first place.

Figure 8.2 Spontaneous mobility-hub-like space (motorcycle taxi station)

Source: (Photo taken by Fumihiko Namakamura)

8.3 TICKETING COORDINATION

Fumihiko Nakamura

8.3.1 Fare Payment in Public Transportation

In emerging countries, not only are there various issues regarding fare payment for public transportation, but there are various technological advancements as well.

8.3.1.1 Passengers' Points of View

From the user's perspective, the issues are the appropriateness of the amount and the ease of understanding the payment procedure. For the appropriateness of the fare, the fare must be reasonable. Regarding payment procedures, for example, if there are various modes of transportation and various operators collect fares independently, even if IC cards and so on are introduced, it will be necessary to prepare different cards for each operator.

8.3.1.2 Operators' Points of View

From the operators' perspective, the issue is reliable fare revenue. There is concern that crew members may commit fraud in the fare collection process. Increasing revenue through various measures may also be an issue.

8.3.1.3 *Policy Framework's Point of View*

From the policy side, from the perspective of inclusive measures, the issue is to consider the poor, while from the perspective of measures to reduce private car use, the issue is to motivate the current private car users.

8.3.1.4 *Fare-Leakage Risk*

If fare collection and management can be separated from the crew, it will be possible to consolidate disparate services provided by different operators and reduce the risk of crew fraud (fare leakage). In Dhaka, Bangladesh, there was a report of a demonstration experiment that showed an increase in revenue by simply changing the bus service to an advance ticket sales system, in which tickets are purchased before boarding and a receipt is issued, but it is said that this does not mean that the number of users increased, but that the number of fraudulent rides decreased.

8.3.1.5 *Preferential Fares*

Once a flat fare structure can be presented, the logic of introducing preferential fares for low-income groups, students, and people with disabilities in the form of discounts or exemptions becomes easier to organize. In Curitiba, the flat fare within the city is not cheap, but those who are subject to the city's welfare policy, such as low-income earners and the elderly, are eligible for a discount from the fare.

Note that in Curitiba, some bus terminals have small supermarkets exclusively for low-income residents, and certified low-income residents can enter the bus terminals and use these facilities. This is known as a complementary attempt to actively connect low-income people with public transportation.

8.3.2 Modernization of Fare Management

8.3.2.1 *Smart Fare Medias*

If new technologies such as IC cards can be introduced in addition to the efforts described in the previous sections, for example, prices can be flexibly adjusted according to time of day, area, and distance traveled, thus realizing a more reasonable fare structure. Management on the operator side would also become more efficient. In emerging countries, there may be no examples of flexible price fluctuation. However, developed countries have introduced fare menus that are possible only with IC cards, such as discounts after 9:00 a.m. and one-day passes in the form of free fares for the third and subsequent times within the same zone on the same day.

Figure 8.3 A gate for Red Line train in Bangkok service with credit card touch transaction

Source: (Photo taken by Fumihiko Nakamura)

The introduction of credit card touch payment is becoming common in emerging countries, and future trends will be watched closely, but the issue will be how to deal with the low-income groups who cannot obtain credit cards as will be mentioned later. In the case of Bangkok, it appears that credit card payment has been introduced for urban railroads with high fares, while buses with low fares, which are supposed to be for low-income passengers, accept cash payment, thus striking a balance between the two.

An increasing number of cities, especially in Europe, are introducing subscription-type fares on smartphone apps for a fixed monthly fee. These so-called "Mobility as a Service" (MaaS) applications are beginning to be deployed in many cities.

8.3.2.2 Mobility as a Service (MaaS)

MaaS is very often understood as an app for public transportation. However, MaaS is essentially a concept, not an app in itself. The correct understanding of MaaS is that it bundles together the various services of various means of transportation and, if necessary, information on commercial facilities, hospitals, and tourist attractions, so that they appear to be a single service to people who are trying to get around.

In addition, information is provided centrally; fares are paid centrally; a monthly subscription system is available; and sharing services, cab services, on-demand services, ride-sharing, and cab reservations can all be made with the same operation. The menu is available. And since there are so many cases where these are realized on smartphone applications, it is often understood that MaaS = smartphone applications. This is fine in terms of the actual situation, but it should be emphasized here that MaaS is originally a concept.

In addition, there are some misconceptions about MaaS, such as that it must include on-demand buses or that it must have self-driving services, but there are no such restrictions at all.

Historically, MaaS was developed as a policy tool to reduce private car use in the urban area of Helsinki, the capital of Nordic Finland. Later, it was also used as a strategy to manage demand in tourist destinations, to disperse demand concentrated at certain times and places, and to promote outings for the elderly and others in depopulated areas. In some regions, such as Japan, the technology has also developed as a tool for enclosing private transportation operators. However, to return to its roots, MaaS can be a tool for reducing the use of private vehicles, and in this context, it is not necessarily limited to developed countries, but can be applied in many situations in large cities in emerging countries.

As can be seen from the discussion so far, the effect of the introduction of MaaS will be that MaaS will change people's transportation behavior. In terms of transportation choice behavior only, it is expected that private car users and motorcycle users will change to public transportation users through MaaS. What we need to be careful about here is the mechanism of this change. For example, it is difficult to imagine that people who stick to using their own cars (private car captives) will download and use apps for public transportation. Even if information is provided in a way that reaches the daily behavior and thinking styles of the private car captive group, thereby changing their perceptions and values, and thus changing their behavior, it is necessary to fully examine whether the information reaches them.

An example of an attempted strategy to put public transportation apps in the hands of the private car captive population is found in Ecuador, South America. One of the country's venture companies is developing a central city parking search assistance app for private car drivers for several South American cities. This company is promoting the development of a MaaS app and is considering strategies to make users of its parking search assistance app aware of public transportation information.

8.3.3 Credit

One point should be noted regarding low-income groups in emerging countries. For example, if the service is limited to the use of smartphones or credit cards, there is always a risk of excluding those who are unable to obtain credit due to their income, that is, those who cannot sign contracts. While it is not necessary to leave the possibility of cash collection, it is necessary to prepare for those who cannot sign up for credit cards or smartphones by using ID cards and so on.

In any case, integrating and upgrading the fare structure of regional public transportation services in this way is one important aspect that must not be forgotten. When introducing urban railroads and BRT, it is important to establish an integrated fare structure for the system that includes the services connected to them, and efforts should not be neglected in this regard.

8.4 FINANCIAL MANAGEMENT

Fumihiko Nakamura

8.4.1 Why Is Financial Management Needed?

Prior to the 1980s, transportation services provided by lively small buses and small cabs, using small converted vehicles and, so to speak, the wisdom of life, met the needs of the community in a meticulous manner. In those days, local public transportation services could be established without any particular financial management. Even today, this situation remains the same for small communities.

Even in large cities, bus services that can attract a reasonable number of users should be able to cover their costs with fare revenues. However, if fares are kept low and uniform for political reasons, it will be difficult to cover the cost with fare revenues, especially as urban areas expand, routes become longer, and users travel greater distances. In addition, if the so-called head office functions are enhanced in a semi-public or near-public management system, the number of staff engaged in activities other than the actual business will increase, and costs will rise significantly. The combination of these factors will make business management more difficult. When a new track system is built in an existing city, the cost is greatly affected if it involves land acquisition. Even if no land other than garages is acquired and an elevated structure is built on a road, river, or canal, or on top of an existing railroad right-of-way, it will cost a certain amount of money, and even more so if the structure is built underground. If these costs are to be reimbursed by borrowing money, and if political pressure imposes restrictions on fare setting, it will be difficult to manage the project.

In this light, while there is no doubt about the importance of introducing public transportation capable of mass transportation in large cities in emerging countries, where the size of cities is expanding and motorization is progressing, a mechanism for securing financial resources for such transportation needs to be carefully prepared.

8.4.2 Several Strategies for Financial Management

There are three main strategies for financing the development and operation of urban transportation in emerging countries. The first will be free or paid assistance from developed countries. The second, which may seem similar to the first, is support based on transactions with developed countries in the context of environmental strategies, such as the use of the Kyoto Protocol's CDM (Clean Development Mechanism). The third is the concept of utilizing financial resources separate from public transportation, such as gasoline tax revenues within the home country.

8.4.2.1 Aids by Developed Countries

The first type, which requires no special explanation here, includes those provided directly by national governments and those utilizing international financial institutions such as the World Bank, each of which has its own characteristics. It should be noted that the experts of the funding organization are not necessarily familiar with the situation in the target country, nor are they familiar with the target urban transportation system or the urban transportation strategy that is the premise for the transportation system. It is important to develop urban transportation experts in each country.

8.4.2.2 Utilization of Environmental Policy Framework Aids

For the second one, the example of Bogotá, Colombia, is instructive. Bogotá adopted a method called CDM based on the Kyoto Protocol after COP3, and developed an urban transportation system with the financial resources of the donor country under the agreement that the amount of carbon dioxide emissions reduced by the introduction of public transportation could be counted as an achievement of the donor country. In Bogotá, the BRT system replaced conventional minibuses with new articulated buses, which were converted from used trucks and other vehicles, and the reduction in CO_2 emissions was counted as an environmental contribution by the donor. In addition, a ropeway system for urban transportation in Medellín, Colombia, has also received support from Europe based on a similar approach.

8.4.2.3 Special Tax Revenue Strategies

As for the third method, a clear example would be the BRT in Bogotá. The Bogotá Metropolitan Region imposes an additional tax on the purchase price of gasoline. It has been agreed that the tax revenue will be used for urban transportation infrastructure, and is being put into the coffers to finance the construction of BRT infrastructure as well as bicycle paths and bicycle parking infrastructure that are being built at the same time.

8.4.2.4 Summary

The third of the aforementioned methods requires political consensus, and it is premised on the assumption that motorization in the country has progressed to a certain degree, but there is concern that consensus will not be easy to reach, given the fact that automobile use is being restrained by economic means in some respects. The second method requires a process of calculation and approval of CO_2 emission reduction benefits. Since there are not necessarily many cases where CDM could be applied to the development of urban transportation systems, careful preparation is required. If this is the case, it is expected that the traditional support of developed country donors and international financial institutions will continue to be the central method.

8.5 CONCLUSION

Fumihiko Nakamura

As cities grow, the design of public transportation networks becomes increasingly important. Especially in large cities in emerging markets, there are multiple rail and bus lines, and there is a need to create an efficient network that connects them. While railways cover a large area, bus routes reach every corner of the city. In addition, the main line branch line network is a concept that establishes branch lines that cover the surrounding areas based on the main rail and bus lines.

When promoting public transportation planning, it is essential to have a master plan that covers the entire network. As individual route plans are developed, the existence of a master plan will clarify the consistency and division of roles between routes and avoid inconsistencies. Particularly in the case of railways, because they involve high investment and long-term development, without a master plan the network may grow but the network effect may not be expected.

In emerging countries, it is important to avoid long-term planning when forecasting demand and to consider a range of possible outcomes. It is important to design a network plan and then check its validity based on actual demand forecasts.

It is important to coordinate routes operated by different operators. From the user's perspective, problems include inconsistencies in transfers, fares, and timetables. The government should take the lead in coordinating between operators. In particular, solutions to physical barriers (inconvenience of transfers) and fare barriers (different fare systems) need to be found at the master planning stage. In recent years, technological innovations such as credit and debit cards have made fare payment more efficient, and it is hoped that these will be used to improve operational coordination and user convenience.

The effectiveness of a network depends greatly on the functions of the transfer facilities that serve as its hubs. The design of transportation hubs must take into account their relationship to urban functions and their location; particularly in emerging countries, many transfer facilities are linked to markets and commercial facilities. For example, in the city of Curitiba in Brazil, the city hall is located next to the bus station, strengthening the link between urban functions and public transportation.

The relationship with roads is also important; especially when using existing arterial roads, it is necessary to design the transportation hub in collaboration with the roads. It is necessary to have a unique design that takes into account the diversity of transportation modes.

Cost sharing should also be considered, and it is necessary to share costs in multiple ways, such as between public transportation operators, road funding, and the use of city funds.

The concept of "mobility hubs" has gained attention in recent years. These are small-scale transportation hubs that provide car sharing, shared bikes, electric vehicle charging, parcel delivery boxes, cafes, and other facilities. In emerging countries, there are cases where these functions have spontaneously sprung up around bus stops, but there are also problems such as illegal use on the street. It is important to manage public spaces and ensure these functions through appropriate design.

There are many issues and technological innovations related to the payment of public transportation fares. From the user's perspective, it is important that fares are reasonable and that the payment method is easy to understand. From the operator's perspective, the issues are the risk of fraudulent revenue and measures to increase revenue. From a policy perspective, it is important to consider low-income groups and to reduce the use of private cars.

On the fare collection side, the risk of fraud by drivers can be reduced. In Dhaka, for example, a system of pre-purchased tickets and receipts has been shown to reduce fraud and increase revenue.

Preferential fares can also be introduced for low-income earners, students, and people with disabilities. In Curitiba, a system has been set up that allows low-income earners to use a small supermarket in the bus terminal, linking public transportation with social welfare.

The use of new technologies makes it possible to set flexible fares. For example, by using smart cards or smartphone apps, it is possible to set prices according to time of day or area, which also improves the operational efficiency of the operator. Especially in emerging countries, credit card payments and subscription-based tariffs (MaaS) are being introduced.

MaaS is a service that integrates multiple modes of transportation, not only public transportation, but also taxis and car-sharing services. This allows users to use a variety of transportation options with a single app, expanding their options. MaaS has the potential to be used effectively in large cities in emerging countries to reduce the use of private cars.

The introduction of MaaS is expected to expand the user base of public transportation and encourage a shift from private cars and motorcycles to public transportation. However, in order to achieve this change, it is important to change the mindset of private car users and provide them with appropriate information.

REFERENCES

Arif, A. A., Falyani, F. K., Sari, V., & Anggraini, R. I. (2023). JakLingko: The implementation of integrated transportation approach in Jakarta smart city. In *Proceedings of the business innovation and engineering conference (BIEC 2022)* (pp. 310–316). Atlantis Press. doi:10.2991/978-2-494069-35-0_36

TraffiQ, & Deutsche Gesellschaft für Internationale Zusammenarbeit (GIZ) GmbH. (2018). *Public transport integration and transit alliances (Sustainable transport: A sourcebook for policy-makers in developing cities, module 3f)*. GIZ. Retrieved from https://sutp.org/publications/sutp-module-3f-public-transport-integration-and-transit-alliances/

Part III

Urban Dimensions and Transportation Interplay

Planning, Equity, and Environment

Chapter 9

Urban Transportation and Land Use

9.1 RELATIONSHIP BETWEEN LAND USE AND TRANSPORTATION

Mihoko Matsuyuki

9.1.1 Relationship between Land Use and Transportation

In cities, social, cultural, and economic activities take place in separate locations. In these activities, some activities such as commuting and shopping are regular activities, while some such as leisure and healthcare are irregular activities. Both of these activities are related with mobility of passengers. In addition, there are production activities related with manufacturing and distribution. Production activities are related with mobility of freight. As these activities take place in different locations, passenger and freight movements are generated. Therefore, transportation and land use are interrelated (Rodrigue, 2024).

Next, we will take a closer look at the interrelationship between transportation and land use. The structure, capacity, and connectivity of transportation infrastructure affect accessibility. And then, accessibility has an impact on land use, which in turn has an influence on activity patterns. If activity pattern changes, it influences distribution and level of transportation demand, which in turn shape planning, maintenance, and upgrade of transportation infrastructure. In this way, transportation and land use have established an interactive cycle (Giuliano, 1995).

It is difficult to demonstrate precisely which has influenced which, so the interactive relationship between land use and transportation is often referred to as a chicken-and-egg relationship. For example, Newman and Kenworthy (1989) studied 32 cities around the world and showed the relationship between gasoline consumption per capita and population density. This study showed that the higher the population density, the greater the amount of gasoline used per capita. This can be interpreted to mean that the higher the population density, the less likely people are to walk, cycle,

DOI: 10.1201/9781003467403-12

or use public transportation, and the more likely they are to use cars. It can also be interpreted that the more cars are used, the more urban sprawl progresses, and the lower the population density gets. In reality, it is thought that these relationships are progressing simultaneously, and that cities with high population density and low car use, or low population density and high car use, are being formed. In addition, TOD is a concept proposed by Peter Calthorpe (1993), and it aims to reduce the car dependency by developing mixed, high-density land use within walking distance of transportation nodes such as train stations. The idea is that mixed, high-density land use promotes the walking and use of public transportation, but to begin with, high-density concentration of functions is only possible because there is a train station.

9.1.2 Land Use Strategies to Reduce Car Dependency

Land use and transportation are deeply interrelated (Aljoufie et al., 2011). For this reason, many concepts and strategies have been proposed and implemented to reduce dependence on cars and promote the use of sustainable travel modes such as public transportation and walking by changing land use and urban structure. The following are some of the most typical examples: polycentric urban development, growth management, compact city, TOD, and Mixed-Use Development (MUD).

9.1.2.1 Polycentric Urban Development

Polycentric Urban Development refers to a metropolitan area characterized by multiple centers of activity, rather than a single dominant center. This urban structure has several key features and implications. The urban area has multiple centers, and population and employment are distributed across the centers. Each center has a certain degree of autonomy and is equipped with the functions necessary for daily life. These centers are connected to each other and the surrounding regions through public transportation networks. This Polycentric Urban Development approach, where housing, workplaces, and service functions are well balanced within each center, has the potential to shorten commuting times by bringing workplaces and services closer to residential areas (Li et al., 2018; Park et al., 2020). Furthermore, the establishment of public transportation networks between centers can reduce dependence on automobiles. If the area of each center is compact, it may also promote mobility through walking and cycling (Park et al., 2020). However, if the distances between centers are long or if the public transportation between centers is inadequate, a significant number of people may rely on cars for inter-center travel, potentially limiting the reduction in automobile dependency (Park et al., 2020). Examples of such Polycentric

Urban Development include Japan's Metropolitan Area Development Plans and the Randstad region in the Netherlands, which comprises the four major cities of Amsterdam, Rotterdam, The Hague, and Utrecht, along with several smaller cities.

9.1.2.2 Growth Management

In the 1950s, the concept of growth management emerged in the United States as a response to various issues associated with urban sprawl, such as the destruction of natural environments. Urban growth management is an approach that strategically plans and regulates urban expansion and development. Key strategies include establishing urban growth boundaries to limit sprawl and promote inward growth, encouraging infill development to redevelop underutilized or unused land within existing urban areas, and implementing mechanisms such as MUD and TOD.

Studies in the United States have shown that states implementing growth management policies experienced an increase in public transportation usage among commuters, while states without such policies saw a decline. Additionally, regions where growth management was applied achieved higher occupancy rates in existing urban areas by curbing unnecessary new developments (Deal et al., 2009). This has been a major factor in improving public transit usage. Such growth management initiatives are particularly prevalent in certain states, such as Oregon.

9.1.2.3 Compact City

"Compact city" aims to curb urban sprawl, address associated environmental issues, and improve the efficiency of land use. According to an OECD report, the key characteristics of a compact city are dense and proximate development patterns, built-up areas linked by public transportation systems, and accessibility to local services and jobs (OECD, 2012). By reducing the distance between housing, workplaces, and various services, a compact city promotes the use of active transportation such as walking and cycling, shortens travel distances, and encourages the use of public transportation due to the connectivity of built-up areas via transit systems. These factors collectively contribute to reducing automobile dependency. Additionally, compact cities are expected to enhance environmental quality, sustain economic vitality in city center, and promote social equity.

A related concept is the x-minute city, which envisions urban areas where residents can access essential services and amenities within "X minutes" by walking or cycling. Notable examples include the 15-minute city in Paris, France; the 20-minute city in Portland, the United States; and the 20-minute city in Melbourne, Australia.

9.1.2.4 Transit-Oriented Development (TOD)

TOD, proposed by Peter Calthorpe in 1993, is a development approach that establishes high-density, mixed-use land use within walking distance of transportation hubs such as train stations and bus stops, creating a pedestrian-friendly environment. TOD is expected to reduce automobile dependency while promoting walking and the use of public transportation. TOD is often implemented in conjunction with Land Value Capture (LVC) methods, leveraging the increased land value generated by real estate development to fund the construction, maintenance of railways, and infrastructure development in TOD areas.

9.1.2.5 Mixed-Use Development (MUD)

MUD is a form of development that integrates multiple functions, such as residential, commercial, office, cultural, and recreational facilities, within a single area. It emphasizes pedestrian-friendly design through the provision of sidewalks, crosswalks, and bike lanes. Vertical integration is also a key feature, with ground floors often allocated for commercial use and upper floors for residential or office spaces, resulting in high-density development.

Such MUD promotes the use of non-motorized transportation by bringing residential, office, retail, service, and recreational spaces into close proximity, potentially reducing car travel distances. Additionally, in mixed-use neighborhoods, the close proximity of various destinations makes walking and cycling more feasible and convenient. There is also a tendency for public transportation usage to increase in these areas (Spears et al., 2019).

In this way, initiatives to reduce automobile use and promote public transportation through land use strategies are being implemented across various spatial levels, from metropolitan regions to neighborhood level. Typically, these multiple strategies are combined, rather than relying on a single strategy, to transition away from a car-dependent city.

9.1.3 Relationship between Land Use and Transportation in Emerging Countries

In this way, land use and transportation are closely interrelated, and in emerging countries, the urban conditions differ significantly from those in developed countries, leading to substantial differences in the relationship between land use and transportation.

One key difference between developed and emerging countries is the speed of urbanization. While major cities in developed countries took approximately 100 years to grow from a population of 1 million to megacities of 10 million, major cities in emerging Asian nations have reached this population level within just 50–60 years. This rapid population growth in major cities has been driven by rapid economic development. Accompanying this rapid population growth, urban areas have also expanded spatially at a remarkable pace, leading to a rapid acceleration of urban sprawl.

While the rapid population growth and spatial expansion of cities continue, the development of infrastructure, such as roads and public transportation, has not kept pace with the speed of urbanization, resulting in constant capacity shortages. Consequently, severe traffic congestion has become a major issue in many large cities in emerging countries.

Compared to developed countries, the timing of the development of railways is delayed, allowing motorization to advance beforehand. This has often led to a form of urban expansion known as ribbon development, where urban sprawl extends along major arterial roads radiating from city centers to suburban areas. In general, ribbon development is considered inefficient in terms of land use and the provision of utilities such as water and sewage.

The interplay between transportation changes, exemplified by motorization, and land use changes, represented by sprawl, has rapidly progressed, each influencing the other. As a result, urban problems such as traffic congestion and environmental pollution have become increasingly severe.

Furthermore, in many emerging countries, urban rail systems are often introduced after urban areas have already expanded. In such cities, constructing railway lines within existing urbanized areas is often constrained by land availability, necessitating elevated or underground structures.

Additionally, if areas around the stations are already urbanized, it becomes challenging to establish station plazas and secure spaces for transferring between feeder transportation and rail systems. In cases where land has already been subdivided into smaller parcels, implementing TOD becomes significantly more complex.

Moreover, emerging countries often have looser land use regulations compared to developed countries, or even when regulations are rigid, their enforcement is frequently less stringent. This creates conditions that are more conducive to sprawl driven by motorization. In addition, investments in new infrastructure such as roads and transportation systems tend to trigger land and real estate speculation, with investors anticipating sharp increases in land values. Compared to developed countries, such investments in new infrastructure in emerging countries are likely to have a much more significant impact on urban form and land use patterns (Cervero, 2013).

As mentioned earlier, land use conditions and transportation systems in emerging countries differ significantly from those in developed countries, and the interactions between them vary as well. In this chapter, we will present the following case studies that illustrate the relationship between land use and transportation: the extent to which the introduction of railroads has influenced housing development in Bangkok; the impact of the decentralized urban structure, as outlined in Hanoi's urban planning master plan, on people's travel behavior; and TOD along the BRT system in Curitiba.

9.2 THE IMPACT OF URBAN RAILWAYS ON PREVENTING URBAN SPRAWL: CASE OF BANGKOK, THAILAND

Mihoko Matsuyuki

9.2.1 Introduction

This section examines how transportation infrastructure, especially urban railroads, is affecting urban development in metropolitan areas. Currently, new urban railroads are being built or extended in many large cities in emerging countries to address urban problems such as air pollution and traffic congestion caused by motorization and urban sprawl. While it has been argued by many scholars that motorization has led to urban sprawl, it has not been fully discussed how urban railways affect the urban development and whether or not it inhibits urban sprawl.

In this section, we examine how the urban railways has changed urban development trends and whether it has suppressed urban sprawl in Bangkok, Thailand. In Bangkok, urban railways were introduced in late 1999 and the network has been steadily expanded, and the landscape around urban railroad stations has been changed significantly to date.

9.2.2 Development of Urban Railways in Bangkok

9.2.2.1 History of Railway Development in Bangkok

The first urban railway in Bangkok, two lines of elevated railway called Bangkok Transit System (BTS) or Skytrain, was opened in December 1999 by the Bangkok Mass Transit System Public Company (BTSC). Then, one line of subway, called MRT, was opened by the Bangkok Metro Company in July 2004. Subsequently, the Airport Rail Link (ARL), an elevated railway connecting the city center to the new airport by the State Railway of Thailand, opened in 2010. Then, three MRT lines connecting city center to the suburbs started their operation in 2016, 2023, and 2023, respectively, while the guided-rail transit system with unmanned operation by the BTSC have been opened in 2021. In addition, a line connecting the northern part of Bangkok to the north and a line connecting the northern part of Bangkok to the west, operated by the Thai state railway, opened in 2021. As of 2024, 10 urban rail lines are in operation in Bangkok and its neighboring provinces.

9.2.2.2 Suburbanization of Bangkok and Changes in Urban Development Trends with the Operation of the Urban Railways

Although motorization began in Bangkok in the 1960s, the rate of private car ownership rose rapidly in the 1980s as the economy grew. The expansion of Bangkok's urban area also began to expand in the 1960s, but its expansion accelerated after the 1980s, when the national economy also grew significantly.

This expansion of Bangkok's urban area into the periphery is not only due to mixture of housing development, the location of industrial parks, and large-scale shopping centers on the periphery of the city.

Originally, the majority of the housing in the central area of Bangkok was two-three story shophouse or townhouse. As the population moved to the suburbs due to population growth and environmental degradation in central Bangkok, residential housing estate began to be built in the suburbs from the 1960s, especially in the 1970s onward. Since the 1970s, many gated communities consisting of low-rise houses (detached houses and townhouses), called *mubanchatsan*, have been built in the suburbs, and the main buyers of these houses have been the middle class (Askew, 2002). Living in a gated community in the suburbs and driving a car to work in the city center has been the lifestyle that the middle class admired.

The first urban railroad, the BTS, opened at the end of 1999 in the midst of this outward expansion of the urban area in Bangkok. As the number of passengers increased, real estate development around railroad stations boomed. In central Bangkok, shopping centers, high-rise office buildings, and condominiums were developed one after another around railway stations. Even outside the center of Bangkok, shopping centers and condominiums were developed around railway stations. More than 20 years have passed since the operation of the urban railway, and the cityscape of Bangkok and its vicinity, especially around railroad stations, has changed dramatically.

9.2.3 Changes in Residential Development in Bangkok Metropolitan Region

9.2.3.1 Differences in Residential Development Trend before and after Urban Railway Development in Bangkok City

Thus, there has been a boom in the construction of high-rise condominiums around urban railway stations in Bangkok and the five surrounding provinces (Bangkok Metropolitan Region: BMR). This trend seems to be the opposite of the conventional preference of the middle class for suburban single-family homes. It appears as if urban rail has curbed the phenomenon of housing sprawl. Thus, this section examines how housing development has changed along with urban railroads. At first, we examine changes in residential development trends in Bangkok and then, we examine it in BMR.

We used the 2006 and 2013 editions of an electronic map called MapMagic released by Think Net. Co. to identify residential building, condominium/apartment house, and gated community built before 2005 and ones built between 2006 and 2013. The first urban railroad, the BTS, opened in December 1999, but since the oldest version of MapMagic is 2006 and real estate development was staggered from the opening of the railway, we use the 2006 version to understand the situation before the railway operation

(Matsuyuki, 2019). Figure 9.1 shows that the construction of condominiums in Bangkok has increased dramatically since 2006, and most of them are located in or near the city center near the railroad stations.

Next, to examine in which parts of the Bangkok City condominiums/apartments and gated communities have been built, the Bangkok City is divided into three areas: central area, middle area, and suburban area based on Bangkok's zoning regulations as shown in Figure 9.2.

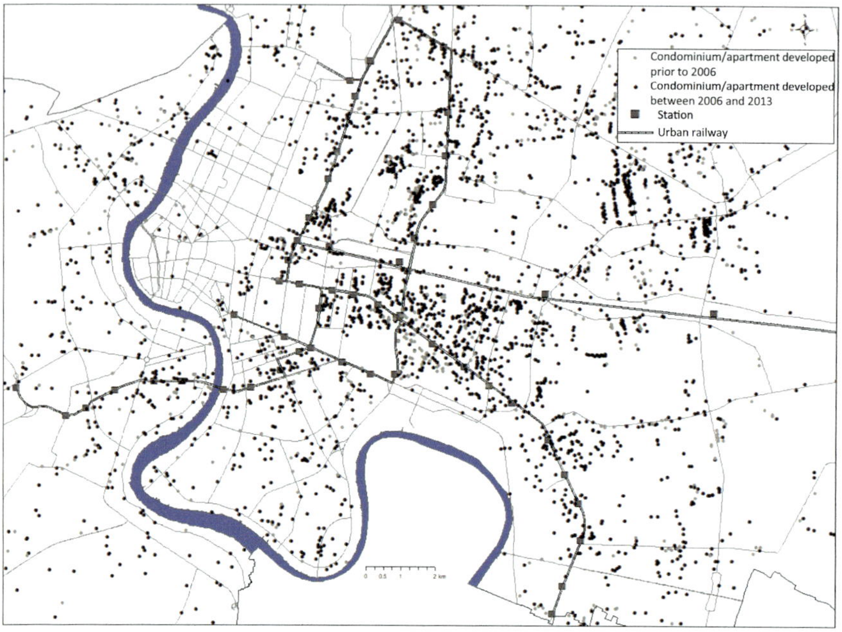

Figure 9.1 Location of condominium/apartment development in center of Bangkok

Source: Matsuyuki (2020)

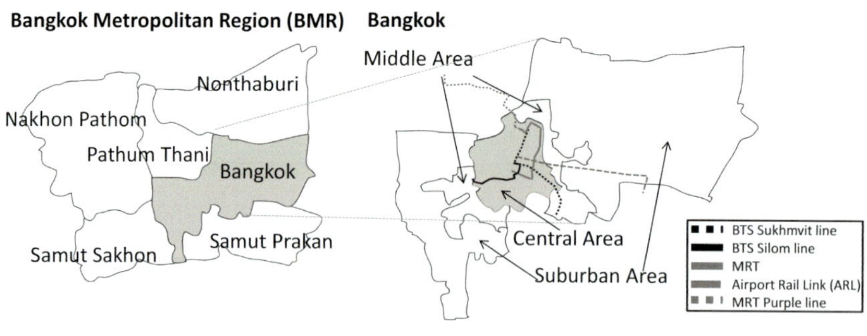

Figure 9.2 Map of BMR and Area Division in Bangkok

* Railroad routes are as of 2013

Table 9.1 Distribution of Residential Development in Bangkok City

		Central area	*Middle area*	*Suburban area*	*Total*
Gated community	Prior to 2006	223 (9.5%)	341 (14.5%)	1,781 (75.9%)	2,345
	Between 2006 and 2013	54 (5.3%)	113 (11.0%)	857 (83.7%)	1,024
Condominium/ apartment	Prior to 2006	322 (41.5%)	148 (19.1%)	306 (39.4%)	776
	Between 2006 and 2013	1,554 (41.4%)	1,022 (27.2%)	1,177 (31.4%)	3,753

Prior to 2006, there were 1,024 gated communities and 776 condominiums/apartment houses in Bangkok, with gated communities predominating. However, 3,753 condominiums were built between 2006 and 2013, compared to only 1,024 gated communities being developed in the same period (Table 9.1).

Table 9.1 shows number of developments in each area in Bangkok city. The share of condominium/apartment construction locations remained almost the same in the central area. This is probably due to the fact that there is not much room for development in the central area. On the one hand, the share of condominiums/apartments built in the suburban area has decreased, while the share of condominiums/apartments built in the middle area has increased. In other words, condominiums are now being developed closer to the city center than in the suburbs. On the other hand, the share of developed gated communities has not changed in the middle area, but the share has increased in the suburban area, meaning that gated communities have spread outward. Since condominium/apartment development has become the mainstream since 2006, overall, residential development in Bangkok is tending to converge inward from the suburban areas toward the middle and central area in Bangkok.

9.2.3.2 *Differences in Residential Development Trend in Bangkok City before and after Urban Railway Development in Bangkok Metropolitan Region*

Next, Table 9.2 summarizes residential development trends in the BMR.

Prior to 2006, 77% of the development was gated communities, with condominiums/apartments accounting for only 23%, indicating that gated community was the main type of housing development. By contrast, since 2006, gated communities and condominiums/apartments have reversed in its share, with gated communities accounting for 32% and condominiums/apartments for 68%, indicating that mid- to high-rise condominium/apartment buildings are becoming the mainstream. In Bangkok, the percentage

Table 9.2 Location of residential development by development period and housing type

		Bangkok	*Nakhon Pathom*	*Nonthaburi*	*Pathumtani*	*Samut Prakarn*	*Samut Sakorn*	*TOTAL*
Gated community	Prior to 2006	2345 (75.1%)	29 (85.3%)	518 (82.1%)	155 (89.1%)	89 (82.4%)	43 (72.9%)	3179 (77.0%)
	Between 2006 and 2013	1024 (21.4%)	123 (53.5%)	368 (50.3%)	550 (51.4%)	357 (46.8%)	85 (43.1%)	2507 (32.3%)
Condominium/ apartment	Prior to 2006	776 (24.9%)	5 (14.7%)	113 (17.9%)	19 (10.9%)	19 (17.6%)	16 (27.1%)	948 (23.0%)
	Between 2006 and 2013	107 (78.6%)	363 (46.5%)	519 (49.7%)	406 (48.6%)	112 (53.2%)	5260 (56.9%)	107 (67.8%)

of condominium/apartment buildings increased rapidly from 25% to 79% prior to and after 2006, indicating that the majority of current residential development is in the form of mid- to high-rise condominium/apartment buildings. In the other provinces, condominium/apartment construction was 10%–30% prior to 2006, while construction after 2006 was around 50%, indicating that the construction of mid- to high-rise multi-family housing is not limited to Bangkok, but also extends to neighboring provinces.

Regarding the location of residential development, prior to 2006, Bangkok accounted for 76%, followed by Nonthaburi Province with 15%, indicating that more than 90% of residential development in the BMR occurred in these two provinces. However, looking at the development after 2006, the shares of Bangkok and Nonthaburi decreased to 62% and 9%, respectively, while those of Pathum Thani and Samut Prakan provinces increased to 14% and 10%, respectively, indicating that residential development is shifting from Bangkok and Nonthaburi provinces to the provinces surrounding northern and southern parts.

Looking at the BMR as a whole, the suburbanization of residential development is progressing from the Bangkok city to the north and south. While looking at only Bangkok, development tends to concentrate from the suburban areas toward the middle and the city center area. In other words, while sprawl toward the outward is progressing in the BMR as a whole, sprawl within Bangkok has stopped. The suppression of sprawl in Bangkok is assumed to be due to the urban rail system, but even so, its impact is limited to Bangkok and has not been able to influence the suppression of sprawl in the BMR as a whole (Matsuyuki, 2019).

9.2.4 Conclusion

In BMR, urban railway has suppressed residential sprawling only in Bangkok, and residential sprawling in the BMR has continued. To suppress sprawling in BMR, the expansion of rail network to spread throughout the BMR is essential.

One of the reasons why the urban railways had such a significant impact on urban development in Bangkok is that the target of the urban railway is the middle class. Although the train fare is more expensive than buses, it does not get stuck in traffic jams, and trains are more comfortable and look more modern than buses. So low-income people who want to save on fares continue to use buses, while the middle class, who value punctuality and speed over getting stuck in traffic, prefer to use urban railways. In other words, whether urban railways have an impact on urban development, and in particular, on sprawl, depends on the middle class, who can afford to own cars. Whether urban railways can prevent urban sprawl depends on whether the middle class will come to prefer using urban railways, dispelling the perception that riding in a car is a sign of social status and that public transportation is a vehicle for the lower classes.

9.3 EVALUATING THE EFFECTIVENESS OF POLYCENTRIC URBAN STRUCTURE WITH MIXED-USE DEVELOPMENT IN REDUCING AUTOMOBILE DEPENDENCY: CASE STUDY IN HANOI, VIETNAM

Mihoko Matsuyuki

9.3.1 Introduction

This section examines how urban planning and urban development may affect travel behavior. In Hanoi, the capital city of Vietnam, a polycentric urban structure is being promoted through a urban planning master plan, with large-scale mixed-use developments, including residential complexes, being advanced in suburban cores. This section examines how such land use strategies, which are considered to mitigate automobile dependency, influence the travel behavior of residents in an emerging country.

9.3.2 Case Study of Hanoi, Vietnam

9.3.2.1 Land Use and Transportation Network in Hanoi

Hanoi merged with neighboring districts and province in 2008, resulting in a significant increase in its city area and population. This led to a significant change in real estate development trend in Hanoi. Since then, many high-rise buildings have been constructed and a modern urban area has been developed.

One of the spatial characteristics of Hanoi is its Old Quatre. The Old Quarter has been the center of the city, and many historical buildings from the French colonial period remain there. Therefore, it is designated as a historical district, and the development of high-rise buildings is strictly restricted. Hanoi has four Ring Roads from Route 1 to Route 4, and the Old Quarter is located inside Route 1. The area between Ring Roads No. 1 and No. 2 was developed in the 1970s, the area between No. 2 and No. 3 was developed in the 2000s, and the area between No. 3 and No. 4 and the area outside No. 4 have been developed after that. Thus, the development of the urban area is expanding outward from the Old Quatre year by year (Figure 9.3).

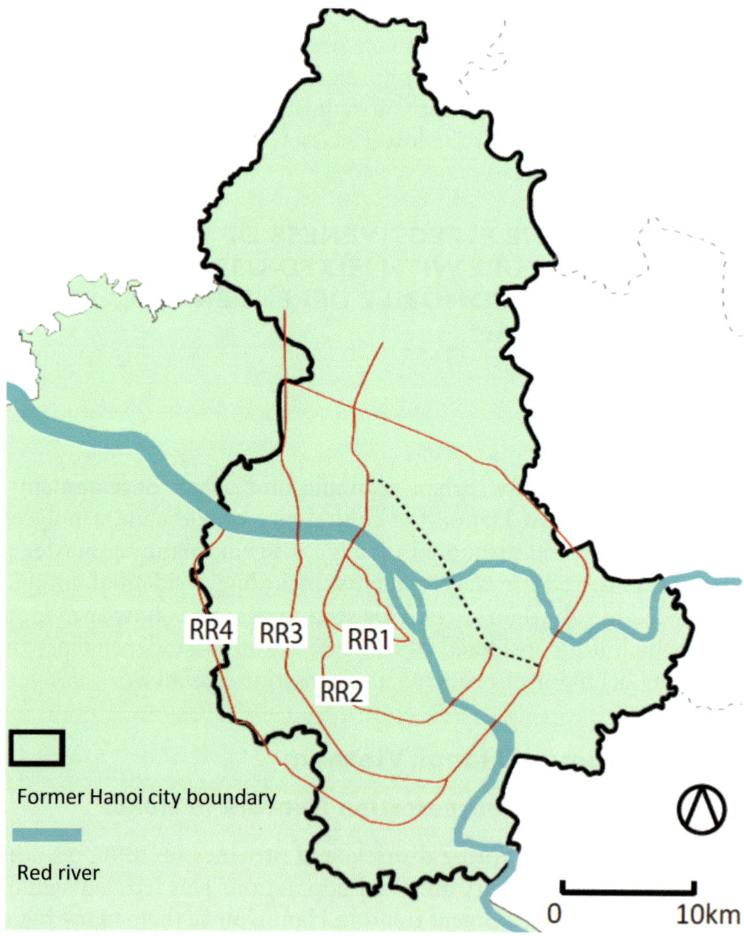

Figure 9.3 Four Ring Roads in Hanoi

Figure 9.4 Tube house (left) and mixed residential development (right) in Hanoi

Source: (Photo taken by Mihoko Matsuyuki)

There has been no mass transit but Hanoi Metro Line 2A opened as the first mass transit in 2021, followed by Line 3 in 2024, and the construction of Lines 1 through 8 is planned in the future. Other public transportation includes buses and BRT, but modal share of bus is not high, at 12.2% of the total. Motorcycles are the main form of transportation in Hanoi, accounting for 64.4% of modal share. The share of car is 2.2%, but the car ownership rate is increasing year by year, and the share of car is also increasing year by year (Each figure of modal share is as of 2020) (JICA, 2020).

One of the reasons for the overwhelming share of motorcycles in Hanoi is the street network and housing in the Old Quarter. Many of the streets in the Old Quarter are narrow, and motorcycle is the most suitable mode to maneuver the narrow alleys and get around. In addition, in the Old Quarter, there are many houses called "tube houses," which were first built in Vietnam in the 19th century. As the property tax rate was once determined by the length of the frontage, the frontage of the building is very narrow (2–5 m) but the depth of the building is long, several tens of meters. The height of the buildings is two stories in the older ones, but some of the newer ones are over 10 stories high. There is no parking space for a car in tube houses, and motorcycles are parked in the narrow alleys between tube houses or inside the houses. These narrow streets and lack of parking space prevent transition from motorcycles to cars, and it is one of the factors in the high motorcycle share.

9.3.2.2 *Urban Planning Master Plan in Hanoi*

" Hanoi Capital Construction Master Plan to 2030 and Vision to 2050" was approved by the Prime Minister in July 2011, as an urban planning master plan which describes the comprehensive development plan up to 2030 and the comprehensive development strategy up to 2050. It describes a vision for the future and the city's vision, as well as land use and other policies to realize these visions. The plan envisions a multicore city with eight core areas, including a central urban area, five satellite cities, and three eco-cities. The urban center area inside the 2nd Ring Road is designated as a historical area, the area between the 2nd and 3rd Ring Roads is designated as an urban improvement area, and the area between the 3rd and 4th Ring Roads and area north of the Red River is designated as an urban expansion area. In the historical area, the construction of high-rise buildings is restricted, but in the urban improvement area, the restrictions are relaxed only along the Ring Road and radial road, and high-rise buildings can be constructed. In urban expansion areas, high-rise buildings can be constructed throughout the entire area. High-rise buildings can be constructed in satellite-cities, although building height and density regulations differ from city to city. Thus, Hanoi's master plan is designed to encourage development by loosening building height and density regulations as it moves outward from the historic area.

9.3.2.3 *Large-Scale Mixed-Use Development*

As described earlier, Hanoi's master plan outlines urban expansion, and residential development took place outside the center of the city. Since 2010, there has been more housing development outside of Ring Road 3. Especially since the 2000s, mid- to large-scale mixed-development – where one or various types of housing such as condominiums, villas, and tube houses, are built on the same site with commercial facilities, restaurants, and other facilities such as private schools – has been developed in suburbs. In many cases, bicycle and parking lots are provided in the basement of the buildings. In recent years, the number of such mixed-use developments have been increased outside of city center in Hanoi.

9.3.3 Survey about Residents' Travel Behavior and Its Result

9.3.3.1 *Outline of the Survey and Characteristics of the Respondents*

We conducted a survey to the condominium residents within six mixed-use residential development sites in 2016. Of the six developments sites, two were located between Ring Roads 2 and 3, two were located between Ring Roads 3 and 4, and two were located outside of Ring Road 4. Of 203 households responded, 57% were in the 30s, most with a college degree or higher, that is higher education. An average household size was 3.2 members and many of them were nuclear families, which is in contrast to traditional

Vietnamese family style, multi-generation family. As described in Table 9.3, majority of respondents work between Ring Roads 1 and 2, and between Ring Roads 2 and 3, which has a new city center with tall buildings (Ishikawa and Matsuyuki, 2017).

9.3.3.2 Changes in Housing Types and Location

The type and location of previous house is described in Table 9.4 and Table 9.5, respectively. Majority (66.2%) had previously lived in a tube house. It indicates that most of respondents have changed their housing type by moving to a mixed-use residential development site. We also asked locational change of residence (Table 9.4). 61.4% of respondents said that their move to the current house made them further away from the city center, while 12.4% said that their move made them closer (Table 9.5). It indicates that the increase in mixed-use residential developments in the suburbs is contributing to the suburbanization of settlement.

9.3.4 Changes in Vehicle Ownership and Travel Behavior

9.3.4.1 Changes in Vehicle Ownership

Next, we examine how motorcycle and car ownership were changed before and after residential moving.

Table 9.3 Location of work place

Inside Ring Road 1	0.8%
Between Ring Roads 1 and 2	35.3%
Between Ring Roads 2 and 3	31.2%
Between Ring Roads 3 and 4	24.1%
Outside Ring Road 4	8.6%

Table 9.4 Type of previous house

Tube house	66.2%
Apartment	21.6%
Condominium	7.4%
Single house	2.5%

Table 9.5 Location of previous house

Further from the city center	61.4%
No change	26.2%
Closer to the city center	12.4%

Table 9.6 Number of motorcycles and car ownership

	Motorcycle					Car			
	Average	*0*	*1*	*2*	*More than 3*	*Average*	*0*	*1*	*2*
Previous location	2.0	0.5%	22.2%	56.2%	21.2%	0.3	72.4%	26.6%	1.0%
Current Location	1.7	2.0%	34.5%	56.7%	6.9%	0.4	58.1%	40.4%	1.5%

Table 9.6 shows the average number of motorcycles and cars and distribution of number of owned motorcycles and cars that a household owns before and after moving into the present residence. On the one hand, that average number of motorcycles dropped from 2.0 to 1.7 because the ratio of motorcycle ownership was almost the same, 99.5% in the previous house and 98.3% in the current house, but some households reduced the number of motorcycles. On the other hand, the average number of cars increased from 0.3 to 0.4 and the car ownership ratio increased from 27.6% to 41.9%.

The most common reason for the reduction in the number of motorcycles was the high cost of parking. They parked their motorcycle inside their home, inside their home plot, or on the roadside for free before, but now they need to pay monthly parking fee to park in a parking lot in the development sites. The most significant reason for the new ownership of private cars was the increase in commuting distance, followed by the availability of parking facility. It indicates that the availability of paid parking facility within the development sites is related to the number of motorcycles and cars.

9.3.4.2 *Changes in Travel Modes and Distance for Commuting and Grocery Shopping*

Next, we examine how travel mode for commuting and grocery shopping were changed by moving to suburban mixed-use residential development sites (Table 9.7).

In terms of commuting travel, 22.7% of all respondents reported that they changed their travel mode. The modal share of motorcycles decreased from 80.8% to 69.3%, while cars and walking increased from 16.0% to 24.6% and from 0.5% to 3.5%, respectively. The overall change was in a more environmentally burdensome direction.

For grocery shopping, 49.7% changed their travel mode. Walking rate increased significantly from 33.7% to 75.9%, while that of motorcycle decreased significantly from 61.0% to 21.9%. The average travel distance

Table 9.7 Change of travel mode for commuting and grocery shopping

Travel mode	*Commuting*		*Grocery shopping*	
	Previous	*Current*	*Previous*	*Current*
On foot	0.5%	3.5%	33.7%	75.9%
Bicycle	1.1%	1.1%	4.3%	1.1%
Electric scooters	0.5%	1.1%	0.0%	0.0%
Motorcycle	80.8%	69.3%	61.0%	21.9%
Car	16.0%	24.6%	1.1%	1.1%
Bus	1.1%	0.3%	0.0%	0.0%
Other	0.0%	0.3%	0.0%	0.0%

also decreased from 1.6 km to 0.5 km, which is related to the fact that 95.5% of households use shopping facilities in the development site where they live.

9.3.5 Conclusion

The results of this survey indicate that in Hanoi's polycentric urban structure, where suburban cores feature mixed-use developments, residents' travel behavior diverges significantly between commuting and shopping trips. For shopping, distances became shorter, and walking accounted for the majority of trips. However, for commuting, many residents travel to the central urban area, leading to longer distances and increased automobile use. Furthermore, the move from traditional tube houses to large-scale mixed-use developments with parking facilities has accelerated car ownership among residents (Ishikawa and Matsuyuki, 2017).

Although mixed-use developments and polycentric urban structures are generally believed to reduce automobile dependency, this is typically the case when workplaces are located in the suburban core areas and when robust public transportation links exist between urban centers and suburban cores. In emerging economies like Vietnam, where these conditions are not yet met, such strategies can inadvertently promote automobile use. The findings also reveal that the provision of parking facilities in residential areas encourages car ownership. In emerging cities like Hanoi, where traditional housing that does not accommodate automobiles is still prevalent, the modernization of housing with parking facilities significantly influences motorization (Ishikawa and Matsuyuki, 2017).

These findings suggest that simply adopting land use strategies generally believed to reduce automobile dependency is insufficient. Instead, it is essential to simultaneously implement other measures, including the development of public transportation networks, the decentralization of workplaces, and strategies tailored to the specific characteristics of cities in emerging countries.

9.4 TRANSIT-ORIENTED DEVELOPMENT – CURITIBA CASE STUDY

Fumihiko Nakamura

9.4.1 Definition of TOD

The term "TOD" is a technical term that began to spread from the United States after the 1990s (ADB, 2019). It was originally intended to express the idea that urban development was auto-oriented, that is, aimed at people who own and use their own cars to get around, while it should be redefined as public transportation on regular routes in the city, or transit as it is called in the United States. In TOD, people are supposed to move around the city. Thus, in TOD, people are expected to use public transportation rather than automobiles (Ollivier et al., 2020).

Examples of developments that expected people to use public transportation existed long before the term TOD was first used. A typical example is the development of large housing complexes in front of stations along private rail lines in major Japanese cities. In this section, we will look at the example of Curitiba, Brazil, which was also practicing TOD long before the term was used, and will learn the significance of this development (ITDP, 2017).

9.4.2 TOD in Curitiba

The city of Curitiba, mentioned in Chapter 2, Section 4, established a structural axis in a radial direction from the central district, based on the premise of an excellent bus system. The structural axis was 100–200 m wide (two blocks) and about 10 km long, and was set in four directions: east, west, south, north, and south. Using a method similar to the land readjustment method, sufficient land for roads was secured by reducing the residential land subdivision by 40%. The land for roads was allocated for three parallel multi-lane roads running through the axis, each about 10 km long. The two roads on either side of the axis would be one-way arterials, each designed for a reasonable volume and speed of traffic. The other road, which would be the spine of the axis, would be a four-lane road, with the middle two lanes dedicated to buses. One lane on each side will be reserved for intra-district automobile traffic based on roadside access, which is expected to have low traffic volumes and low travel speeds. While some may argue that there is no need for a dedicated bus lane, a bus-only curb lane has been implemented to ensure the reliable and large number of bus vehicles and to indicate their presence.

On both sides of this road, high-rise residential buildings are being constructed. Since the construction of high-rise residential buildings is prohibited

Figure 9.5 Structural Axis Corridor in Curitiba

Source: (Photo taken by Fumihiko Nakamura)

outside the central area, the development pressure generated by population and economic growth will be concentrated on this structural axis. In addition, the first and second floors of high-rise residential buildings are limited to commercial or business functions. As a result, high-rise residential buildings will be located along bus routes, with shops, restaurants, and cafes on the first and second floors. In addition, due to the low volume and slow speed of traffic in the general traffic lane, it is easy to access the bus stop on the bus lane from the street and cross to the shops on the other side of the street. The design of the project allows residents to live along the roadside where they can enjoy their daily lives and when they need to go to the central area for work and so on, they can simply take a bus that runs on the bus lane in front of them, thus realizing a life without using a car. This example was later presented to the world as Curitiba's TOD.

9.4.3 Current Situation of TOD in Curitiba

As mentioned in Chapter 2, there was a period of time in Curitiba, from 2004 to about 2017, when the policy shifted in the opposite direction of the previous policy. At that time, there was a political movement to use federal subsidies to implement a metro system instead of buses, and bus service on bus-only streets was not always adequate in some respects. In addition, the number of households with cars increased, the space of the structural axis and the attractiveness of high-rise residential buildings caused land prices to rise, and the area along the bus-only road was inhabited by middle- to high-income families who expected to use their own cars. As a result, lower-income residents who had no choice but to use the bus were pushed to the outskirts of the city. Fortunately for them, the flat bus fares did not increase the financial burden of commuting from the outskirts of the city to the central or industrial districts.

As a result of the above, in recent years in Curitiba, the buses on the structural axis have become very crowded, because the residents on the structural axis do not use the buses very often, while many residents, especially low-income people, use buses from outside the city. The distance traveled by users increases, while the flat fare reduces the amount paid per distance traveled. Even though the system is not independently financed, it can be said that this is an undesirable situation from a management point of view (Nakamura et al., 2016, 2017).

9.4.4 Implication with Developments around Railway Stations in Emerging Countries

It is not difficult to imagine low-income residents being displaced as TOD areas become more attractive in terms of property values. The TOD development area may be a bit forced, but it will be a good place to start. Although this may be a bit aggressive, it may be necessary to provide low-income residents with places to live, such as introducing affordable housing in TOD developments.

In Bangkok, Thailand, residents of TODs in front of elevated train stations are said to have chosen to live there not because they want to use the train, but because of the attractiveness of the TOD. In some cases, developers have anticipated this and built apartment complexes in front of the station, but have not taken into account the walking distance to the station.

In light of these examples, we must first ask who is promoting TOD in the first place and for what purpose, whether policymakers have an intention to reduce automobile use or not, and if so, the development strategy must fully consider the mechanism that allows automobile users to conveniently change their mode of transportation. Here it is necessary to classify car users in more detail. The menu of strategies differs between those who use automobiles to commute to work or school and those who come to commercial facilities to shop.

For example, in Portland, Oregon, a leading example of TOD, the strategy is divided between infill TOD at LRT stations close to downtown and newly developed TOD at LRT stations in more outlying areas. The former aims to encourage the use of LRT for both commuting to and from downtown and for shopping, while the outer LRT stations include commercial facilities and encourage car trips for shopping. The project aims to get people to use the LRT for commuting to and from downtown.

To reduce the use of cars for commuting to and from work and school, TOD in conjunction with rail or BRT development in large cities in emerging countries should be combined with the development of safe, all-weather pedestrian routes between residences and stations, ensuring transit capacity during commuting hours, directly connecting offices and, if possible, schools to or near stations, and strengthening management of car parking at offices and schools (e.g., prohibiting or charging high fees for short stops for parking and pickup). On the one hand, it would be necessary to combine these measures with tighter control of car parking at offices and schools (e.g., prohibiting or charging high fees for parking and short stops for picking up and dropping off children). On the other hand, of course, rents and fares also need to be considered from the perspective of ensuring low-income people's access to public transportation.

9.5 CONCLUSION

Mihoko Matsuyuki

This chapter first discussed the relationship between land use and transportation in Section 8.1, emphasizing their mutual influence. The relationship was likened to the "chicken-and-egg" problem, as it is unclear which drives the other. Then, land use strategies aimed at reducing car dependency, such as Polycentric Urban Development, Growth Management, Compact Cities, TOD, and MUD, were then introduced.

In Section 8.2, the impact of transportation infrastructure on land use was examined by analyzing changes in the type and location of residential development before and after the introduction of urban railways in Bangkok. The findings revealed that urban railways curbed sprawl within Bangkok, concentrating residential development around railway stations. However, suburban sprawl persisted across the BMR, indicating that the impact of urban rail on curbing sprawl was limited to the city level.

Section 8.3 explored how the Polycentric Urban Structure and MUD in Hanoi influenced the travel behavior of residents. While the travel distances for shopping trips were shortened and walking became the dominant mode of transportation, travel distances and car usage for commuting increased. Furthermore, the transition to MUD significantly accelerated car ownership among residents.

Section 8.4 analyzed the effects of Curitiba's TOD policies, revealing both successes and challenges. Although the TOD strategy initially improved BRT convenience, increased land values along TOD corridors attracted higher-income residents who rarely used public transportation, thereby diminishing equity and the bus system's efficiency and pushing low-income residents to the urban periphery.

These findings confirm that land use and transportation are indeed mutually influential, forming a dynamic relationship that continuously evolves over time. In the case of Curitiba, which adopted a TOD strategy, the improved convenience of the bus system led to increased land values along TOD corridors. Over time, higher-income residents, who are less likely to use public transportation, moved into TOD areas. This demonstrates that land use and transportation not only influence each other but also undergo significant changes over the long term. Consequently, it is crucial to evaluate these impacts from a long-term perspective rather than focusing solely on short-term outcomes. Moreover, the interplay between transportation and land use is also influenced by external factors such as policy changes, societal dynamics, and changes in economic conditions. To ensure sustainable development, it is essential to monitor changes in transportation and land use over time and intervene when necessary to steer developments in a desirable direction. In emerging economies, where socio-economic conditions and policies can change rapidly, continuous observation and timely adjustments are particularly critical.

Importantly, the introduction of transportation infrastructure does not immediately prevent sprawl, nor does the adoption of land use strategies automatically reduce car dependency. For instance, in Hanoi, the impact of land use strategies on travel behavior differed depending on travel purposes, and in some cases, even promoted car ownership. When implementing transportation and land use strategies, it is crucial to identify the factors driving sprawl and motorization in each country's/city's context to ensure that the adopted measures do not inadvertently exacerbate these issues. This is particularly important in emerging economies, where the mechanisms driving sprawl and motorization often differ from those in developed countries.

The impact of Bangkok's urban rail system on residential development also highlights the need to consider the scope and target of transportation and land use strategies. While urban rail led to a concentration of housing development within the city, residential sprawl persisted in the BMR. This indicates that understanding the range and scale of such impacts is vital for effective policy implementation.

In conclusion, the relationship between transportation and land use is highly complex, with unique mechanisms often being present in emerging economies. To effectively curb urban sprawl and reduce car dependency, it is essential to understand the specific dynamics of land use and transportation in each city's context, especially in emerging countries. Comprehensive and long-term policies, tailored to the characteristics of each city, are indispensable for achieving sustainable development.

REFERENCES

Aljoufie, M., Zuidgeest, M., Brussel, M., & Maarseveen, M. F. A. M. (2011). Urban growth and transport: Understanding the spatial-temporal relationship. *Urban Transport XVII*. doi:10.2495/UT110271

Asian Development Bank (ADB). (2019). *Transit-oriented development policies and station area development in Asian cities*. Author.

Askew, M. (2002). Bangkok: Place, practice and representation (1st ed.). Routledge.

Calthorpe, P. (1993). *The next American metropolis: Ecology, community, and the American dream*. Princeton Architectural Press.

Cervero, R. (2013). Linking urban transport and land use in developing countries. *Journal of Transport and Land Use*, 6(1), 7–24. doi:10.5198/jtlu.v6i1.425

Deal, B., et al. (2009). Growth management and sustainable transport: Do growth management policies promote transit use? *Journal of Public Transportation*, *12*(4), 21–40. doi:10.5038/2375.0901.12.4.2

Giuliano, G. (1995). Land use impacts of transportation investments: Highway and transit. In S. Hanson (Ed.), *The geography of urban transportation* (pp. 305–329). New York: The Guilford Press.

Institute for Transportation and Development Policy (ITDP). (2017). *TOD standard*. Author.

Ishikawa, A., & Matsuyuki, M. (2017). The effect of modern housing development in suburban areas on travel behavior in developing countries: A case study of Hanoi, Vietnam. *Journal of the Eastern Asia Society for Transportation Studies*, *12*, 1058–1075. doi:10.11175/easts.12.1058

Japan International Cooperation Agency (JICA). (2020). *Preparatory survey on Hanoi public transportation management and operation improvement project (First Phase) in Socialist Republic of Vietnam: Final report*. JICA. Retrieved from https://openjicareport.jica.go.jp/pdf/12365391.pdf

Li, X., Mou, Y., Wang, H., Yin, C., & He, Q. (2018). How does polycentric urban form affect urban commuting? Quantitative measurement using geographical big data of 100 cities in China. *Sustainability*, *10*(12), 4566. doi:10.3390/su10124566

Matsuyuki, M. (2019). Difference of housing development and housing preference between before and after urban rail transit development in Bangkok metropolitan region, Thailand. *Journal of Architecture and Planning*, *84*(755), 137–145. doi:10.3130/aija.84.137

Matsuyuki, M. (2020). Station area development in Bangkok: Is Bangkok's station area development transit-oriented development? *City Planning Review*, *69*(5), 94–97. (in Japanese)

Nakamura, F., Toyama, Y., & Makimura, K. (2016). *Bus changes city*. IBS Shuppan. (in Japanese)

Nakamura, F., Toyama, Y., & Makimura, K. (2017). Perspective on an urban transportation strategy with BRT for developing cities. *Engineering and Applied Science Research*, *44*(3), 196–201.

Newman, P., & Kenworthy, J. (1989). Gasoline consumption and cities: A comparison of U.S. cities with a global survey and its implications. *Journal of the American Planning Association*, *55*(1), 24–37. doi:10.1080/01944368908975398

OECD. (2012). *Compact city policies: A comparative assessment. OECD green growth studies*. OECD Publishing. doi:10.1787/9789264167865-en

Ollivier, G. P., Unal, B., & Gunes, B. (2020). *Transit-oriented development for sustainable cities*. World Bank.

Park, K., Ewing, R., Sabouri, S., Choi, D.-A., Hamidi, S., & Tian, G. (2020). Guidelines for a polycentric region to reduce vehicle use and increase walking and transit use. *Journal of the American Planning Association*, *86*(2), 236–249. doi:10.1080/01944363.2019.1692690

Rodrigue, J.-P. (2024). *The geography of transport systems* (6th ed.). Routledge.

Spears, S., Boarnet, M. G., Handy, S. L., & Rodier, C. J. (2019). *Impacts of land-use mix on passenger vehicle use and greenhouse gas emissions* (Brief UCD-ITS-RP-14–136). Institute of Transportation Studies, University of California, Davis.

Chapter 10

Urban Transportation and Equity

10.1 WHAT IS TRANSPORTATION EQUITY?

Mihoko Matsuyuki

10.1.1 What Is Transportation Equity?

There are two major perspectives on transportation equity: horizontal equity and vertical equity (Litman, 2007). Horizontal equity, on the one hand, assumes that individuals with similar needs and abilities should receive comparable shares of public resources and bear similar costs. Vertical equity, on the other hand, focuses on allocating more resources and support to people who are socially and economically disadvantaged. This perspective emphasizes addressing differences in "need" and "ability" by favoring vulnerable groups. Although these two perspectives may seem to involve conflicting principles, transportation planning should treat them as complementary rather than contradictory, adopting an integrated approach to balance both forms of equity. Next, horizontal equity and vertical equity in urban transport are outlined.

10.1.1.1 Horizontal Equity

Urban transportation has a significant impact on our lives. This is because urban transportation services have a major impact on whether we can go to buy healthy food, find work, commute to work, go to school, or access necessary medical care. In addition to these positive impacts, urban transportation also has negative externalities such as traffic congestion, air pollution, and traffic accidents, as well as relocation and gentrification associated with the development of transportation infrastructure (Zuk et al., 2018). Therefore, how the benefits and burdens of such a transportation system should be distributed in society is a very important issue in transportation policy. In reality, in many cases, the benefits and burdens of transportation are distributed unfairly. For example, roads built with a lot of investment are mainly used by people who can use private cars, but people who wait for public

DOI: 10.1201/9781003467403-13

transportation or travel on foot or by bicycle suffer health damage from air pollution caused by substances emitted by cars.

So, how should the benefits and burdens of transportation be distributed? First, the benefits of transportation that should be distributed are mainly accessibility, that is, the ease of getting to major destinations such as places of employment, medical facilities, and educational institutions. In cities where cars are the main means of transportation, the emphasis is more on reducing travel time, that is, easing congestion, rather than on accessibility. Furthermore, the negative externalities of transportation to be allocated include mainly health-related concerns due to air pollution and traffic accidents. Historically, on the one hand, these negative externalities have been unfairly allocated to low-income and minorities who are exposed to high-traffic roads. On the other hand, high-income people who use private cars are not exposed to these risks as much (Schweitzer and Valenzuela, 2004).

10.1.1.2 Vertical Equity

The goals of vertical equity include inclusivity, affordability, and social justice.

Inclusivity aims to ensure that everyone has access to transportation options and can reach essential services and activities. This goal emphasizes accommodating individuals with physical or cognitive limitations and addressing the specific travel needs of groups such as seniors, people with disabilities, pregnant women, and parents traveling with children. Achieving inclusivity requires implementing universal design principles, such as installing elevators at stations, adopting low-floor buses, and providing wheelchair-accessible spaces on vehicles. These measures ensure that transportation infrastructure and vehicles are barrier-free and accessible to all users equally.

Affordability focuses on ensuring that transportation costs do not impose a financial burden, particularly on low-income households. A critical benchmark is that transportation expenses should not exceed a certain proportion of household income. To achieve affordability, measures such as discounted transit fares for students, seniors, and people with disabilities; reduced fares or free passes for low-income individuals; and the development of safe, convenient environments for walking and cycling – low-cost travel modes – are essential.

Social justice seeks to address historical and structural inequities, such as racial, gender, and economic disparities, to ensure that all social groups have equal access to transportation opportunities. This includes preventing minority groups and low-income neighborhoods from being disproportionately disadvantaged. Social justice measures include equitable investment in infrastructure development across all regions and communities, ensuring that public transit infrastructure is available in low-income areas, and mitigating the concentration of air pollution and noise caused by highways and transportation facilities in disadvantaged communities. Providing compensation and mitigation measures for affected residents and enabling meaningful participation of socially disadvantaged groups in decision-making processes are critical for achieving this goal.

10.1.2 Disadvantaged Groups

The disadvantaged groups, that is, those who are unequally distributed the benefits and burdens of transportation, are referred so due to low income, ability, and historical discrimination (Lucas et al., 2019). It comprises young people, women, sexual minorities, low-income earners, households with no cars, elderly people, people with disabilities, single-parent households, refugees, racial/ethnic minorities, and so on. This section focuses specifically on low-income individuals, persons with disabilities, older adults, women, and minorities, outlining the transportation-related inequities they face.

10.1.2.1 Low-Income Group

Low-income household often cannot afford private vehicles, relying instead on public transportation, walking, or cycling. They tend to reside in areas with lower land and housing costs, which often coincide with limited public transportation services. Additionally, transportation costs represent a higher percentage of their household expenses, placing significant financial strain on their daily lives.

When low-income households live in urban peripheries or suburban areas to reduce housing costs, their access to employment opportunities, schools, and essential services may be further restricted. Transportation infrastructure investments are often concentrated in central urban areas or major roads, leading to inequitable distribution of benefits. Furthermore, transportation infrastructure improvements have led to gentrification in many parts of the world, where rising property values and an influx of higher-income residents displace original lower-income residents.

Consequently, low-income households face limited access to public transportation, lower service levels, and a disproportionate burden of transportation costs relative to their income. These factors restrict access to better employment and education opportunities, potentially worsening social exclusion. Poor access to healthcare services also increases the risk of deteriorating health conditions (Litman, 2007).

10.1.2.2 Persons with Disabilities

People with disabilities, such as wheelchair users and individuals with visual impairments, often encounter significant challenges and risks when traveling independently. Limited accessibility to stations, along with insufficient implementation of universal design features like tactile paving or barrier-free facilities, further restrict their use of public transportation. As a result, persons with disabilities often rely on more expensive transportation options, such as taxis, leading to higher travel costs.

10.1.2.3 Older Adults

Older adults frequently experience physical limitations or reduced driving ability, making it difficult for them to rely on private vehicles. Declining physical capacities, including mobility issues, impaired vision, and reduced hearing, can also create significant challenges for walking or using public transportation. Consequently, many older adults must rely on others for transportation.

When public transportation is inconvenient or inaccessible for older adults, their access to medical facilities and commercial centers becomes restricted, leading to a decline in their quality of life. This can result in worsening health conditions, increased isolation, and potential mental health issues.

10.1.2.4 Women

Women are often engaged in unpaid and paid caregiving work, such as housework, childcare, and elder care. These responsibilities shape distinct travel patterns, characterized by shorter trips, more frequent off-peak travel, and multi-destination journeys. Women also have lower rates of driving license and private vehicle ownership compared to men, making them more dependent on public transportation and walking (Hanson, 2010).

However, current transportation infrastructure and operations often prioritize male commuting patterns, particularly peak-hour travel associated with wage labor. This mismatch makes existing transportation systems less user-friendly for women. For instance, public transportation schedules align with peak hours for male commuters, while off-peak services are often limited. Poorly maintained sidewalks and insufficient barrier-free designs, such as facilities for wheelchair or stroller users, further restrict women's mobility. Additionally, women are disproportionately affected by safety concerns, such as sexual harassment on public transportation, which limits their freedom to travel, particularly during nighttime or in crowded conditions (Peters, 2013).

10.1.2.5 Minorities

Racial, ethnic, and religious minorities often reside in areas with lower-quality transportation services and infrastructure due to historical or structural factors (Bullard, 2001). In some countries, gentrification and past segregation policies have further restricted their housing and mobility choices. As a result, minorities often depend on public transportation or walking due to economic constraints but live in areas with insufficient access to public transportation, which limits their access to workplaces and educational institutions, thereby perpetuating economic inequality.

Furthermore, minorities often reside in urban areas with high levels of air pollution, exposing them disproportionately to health risks such as asthma and cardiovascular diseases. In the United States, for example, African American and Latino populations report higher rates of transportation-related health issues. Minorities are also frequently excluded from decision-making processes for new transportation or infrastructure projects, often due to language barriers or systemic exclusion. Moreover, transportation infrastructure investment tends to be lower in low-income areas predominantly inhabited by minorities compared to affluent regions.

10.1.3 Transportation Equity in Emerging Countries

Transportation equity has only recently begun to attract attention in emerging countries. However, with motorization progressing rapidly and public transportation service being insufficiently provided, transportation equity is an extremely important issue in emerging countries, where there is a large gap between the rich and the poor (Vasconcellos, 2011). In addition, emerging countries have a different aspect to transportation equity from that of developed countries.

The first difference is that, compared to developed countries, emerging countries have a much larger income gap and their democracy is still in its infancy, so transportation policies tend to favor high-income groups. Even though walking, cycling, and public transportation share the majority of transportation, the limited resources for the transportation system are often allocated to road development. The development of infrastructure for public transportation, walking and cycling, which are used by lower-income groups, tends to be given low priority. As a result, the infrastructure for public transportation, walking and cycling is insufficient, making it inconvenient and unsafe, and low-income people who use these modes of transportation have to spend a lot of time traveling, and are also exposed to more risks. In other words, there is an imbalance in which the benefits of transport are mainly enjoyed by high-income groups, while the externalities of transport are mainly borne by low-income groups.

The externalities of transportation, such as traffic accidents and air pollution, have a greater impact on vulnerable social groups in emerging economies than in developed economies, because regulations and countermeasures against externalities are either not yet in force or are not as effective in emerging economies.

Furthermore, the quantity and quality of public transportation are insufficient, requiring multiple transfers and long waiting times, which not only takes a lot of time to reach your destination, but also makes the cost of travel expensive. The cost of travel can sometimes account for a very high proportion of the income of low-income groups compared to high-income groups.

As public transportation facilities are not well developed, the installation of facilities for the disabled and elderly, such as elevators, in public transportation facilities is often given low priority. In emerging countries, railway construction often takes place after urbanization has advanced significantly, so in many cases railways are constructed as elevated railways or subways. As a result, accessibility to public transportation by the disabled and elderly tends to be lower than that in developed countries.

As you can see, in emerging countries, the issue of transportation equity has a significant impact on vulnerable groups such as the low-income groups, and the situation is more serious than in developed countries. Dealing with this issue in emerging countries is an urgent matter, but is not attracting much attention. However, because the problem is so serious, improving accessibility for disadvantaged groups specifically is expected to greatly improve the issues they face, and this in turn is expected to lead to a betterment of social issues. For example, if we improve accessibility for the poor in terms of jobs, education, and so on, we can expect to greatly improve issues such as poverty and social exclusion.

Against this background, this chapter will introduce the cases of Curitiba and Medellín, which have attempted to improve accessibility for the low-income population and address poverty issues, and the emergence of gentrification as a negative externality due to the development of rail transport in Bangkok and its impact on the transport system.

10.2 TRANSPORTATION PLANNING AND POVERTY REDUCTION IN CURITIBA

Fumihiko Nakamura

10.2.1 Overview of Bus System in Curitiba

The basic form of Curitiba's bus system was outlined in 1979. The IPPUC, the city's urban planning institute, conducts basic planning, and URBS, the city corporation, conducts detailed operation planning, management of transit terminals and other facilities, fare revenue management, operation contracts with operators in the city, payment of contract fees, and day-to-day operation management based on these plans. Actual operations will be conducted by each operator under contract with URBS. All operators are private companies, but since they operate under contract with URBS, each operator is not in the red. All fare revenues are consolidated into URBS, and fares paid by users do not directly go to the operators' staff. The bus route network is hierarchically organized with trunk lines and branch lines. Transfers between bus routes are encouraged, and bus fares are waived for transfers at transfer terminals. The bus route network is also closely linked to urban planning, and major public facilities, including the district office, are located at the transfer bus terminals.

Fares are also to be determined by URBS. Although fares are flat, there are various detailed regulations. One unique feature is that police officers and post office employees who wear uniforms are entitled to free fares. This is one aspect of the system that encourages the use of buses for official business.

10.2.2 Poverty Issues and Bus Management

In relation to income inequality and poverty issues, there are several characteristics of Curitiba's bus service and operation.

10.2.2.1 Considerations in Fares

As mentioned earlier, public facilities are attached to transit bus terminals. Low-income individuals and others who visit public facilities are entitled to a bus ticket for the return trip. There are no small number of users, including retired senior citizens and unemployed people looking for work. Bus fares are also waived for students from low-income families.

10.2.2.2 Establishment of Routes to Poor Areas (South Circular Route)

In the late 1990s, when Cassio Taniguchi was mayor, the city introduced a road for buses to low-income neighborhoods located in a linear pattern in the southern part of the city under high-voltage power lines, and introduced a trunk bus service with the latest bus vehicles and tube bus stops. The project is also undertaking a major transformation of poor-quality housing areas by providing low-income workers with employment-supportive housing and other services. It has become a trunk line with very high ridership throughout the day.

10.2.2.3 Structural Axis and the Problem of Poverty

The structural axis is a space of high-rise residential buildings along a bus route along which trunk buses run. The high quality of the high-rise residential buildings led to higher rents or sales prices, which in turn led to higher land prices and land rent. This has led to an increase in the average income of residents along the structural axis. Low-income residents found it difficult to live along the structural axis, and although they worked in urban and industrial areas, their places of residence were located at the outer edges of the city or outside the city, beyond the structural axis.

Fortunately or unfortunately, the city has become more coordinated with neighboring municipalities beyond the city limits, and bus fares are flat, even to the outskirts of the city. Therefore, for the poor who have been driven out of the city, bus fares remain constant, although the time required to get to their place of work has increased.

As a result, buses traveling on bus-only roads on the structural axis are extremely crowded with poor people commuting from outside the city to the center of the city for transportation. Ironically, however, the residents on the structural axis earn higher income, own their own cars, and do not often use the buses. Not many of them necessarily board from bus stops on bus-only roads. From the perspective of bus operations, this is not a desirable situation because long-distance users are not paying fares that are commensurate with the actual cost of operating the buses.

The situation is expected to change by providing affordable housing on the structural axis, stopping uniform fares, and so on.

10.3 THE EFFECTS OF INSTALLING CABLE CAR SYSTEM IN LOW-INCOME AREA IN MEDELLÍN

Mihoko Matsuyuki

10.3.1 Introduction

Medellín, Colombia, once plagued by violence and drug cartels, has seen significant improvements in urban mobility in low-income area through the introduction of an aerial cable car system called Metrocable (Figure 10.1). The Metrocable was implemented as a part of PUI (Proyecto Urbano Integral), which refers to an integrated urban project aimed at improving infrastructure and social services in low-income areas. The Metrocable aims to address the transportation disadvantages faced by low-income residents in the city's hilly regions. This section explores the impacts of the Metrocable on residents' travel behavior, economic opportunities, and social inclusion, focusing on Comuna 1, one of the poorest areas in Medellín.

Medellín's geography and rapid urban growth in the 20th century led to the development of extensive low-income settlements on the city's steep hillsides. These areas, often isolated from the city center, lacked adequate infrastructure and accessible transportation options. The Metrocable system, inaugurated in 2004 with Line K, aimed to bridge this gap by providing

Figure 10.1 Metrocable

Source: (Photo taken by Mihoko Matsuyuki)

affordable, reliable, and safe transportation from these marginalized areas to the city's main employment and commercial centers.

10.3.2 Metrocable System Overview

Medellín is situated in a valley surrounded by mountains, with the Medellín River flowing through it. The city center has developed along the Medellín River, while the hillsides are home to low-income residential areas. Figure 10.2 shows the public transportation system in Medellín. The metro system runs along the Medellín River, making it difficult for the low-income population in the hillside areas to access the metro system. Consequently, they face challenges in reaching the city center. The Metrocable system integrates seamlessly with Medellín's Metro, allowing passengers to transfer between cable cars, trains, buses, and trams using a single fare system. Line K, the first Metrocable line, connects the densely populated Comuna 1 with the city's metro network, significantly reducing travel time and costs for residents. Subsequent lines further expanded the system's reach to other low-income neighborhoods.

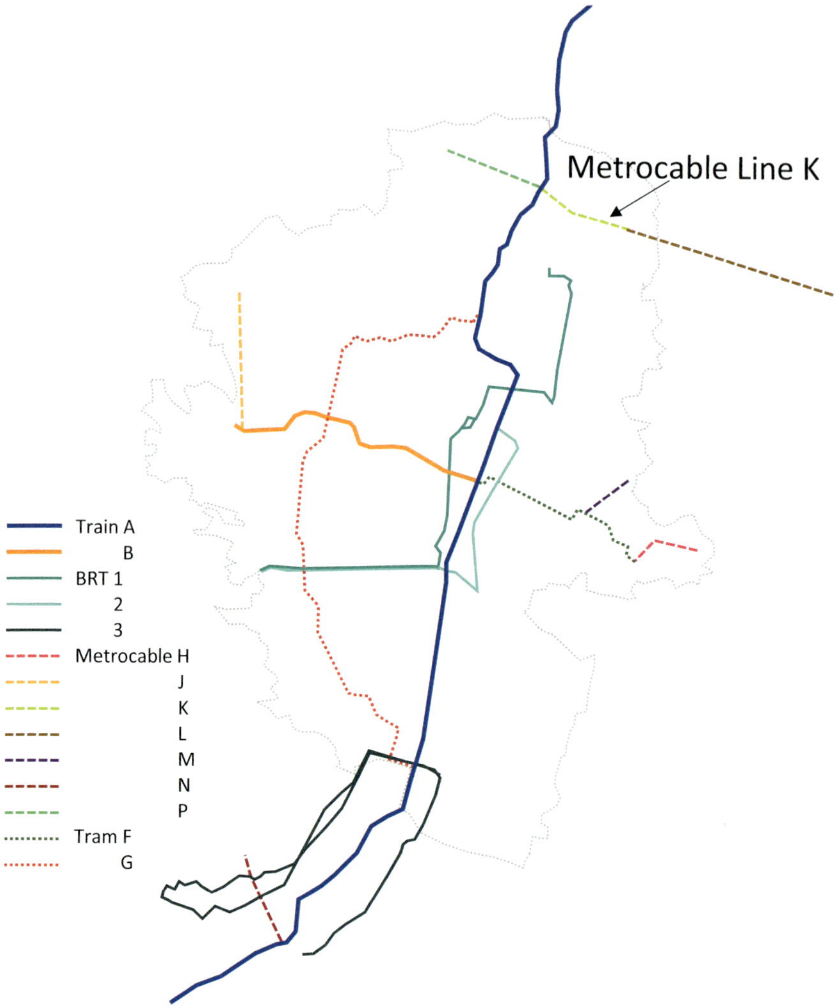

Figure 10.2 Public transportation system in Medellín and Metrocable line K

10.3.3 Survey and Its Key Findings

10.3.3.1 Methodology

To assess the Metrocable's impact on low-income residents' travel behavior and accessibility, and consequently poverty reduction, a survey was conducted among 371 residents of Comuna 1 in which Metrocable Like K runs through, from 2016 to 2017. The survey collected data on residents' socio-economic characteristics, commuting patterns, mode choices, and travel times (Matsuyuki et al., 2020b).

10.3.3.2 Travel Behavior Changes

Table 10.1 shows respondents' commuting travel mode before and after Metrocable operation. Before Metrocable operation, on the one hand, car travel as a passenger (the majority of which is assumed to be informal public transportation) was dominant, where the fraction of car passengers decreased drastically from 48.7% in 2003 to 21.0% in 2016. The second dominant travel mode before Metrocable was bus, where the number of bus users decreased from 37.0% to 26.1%. On the other hand, the fraction of train users increased from 7.6% to 22.7%, which was attributed to the easier access to the train station offered by Metrocable.

The average one-way travel time to work before Metrocable was 59.6 min, while in 2016 it was 44.6 min. This reduction in travel time of about 15.0 min was partially attributed to the Metrocable. We asked respondents why they use Metrocable as a commuting mode, and the most prominent factor to select Metrocable for commuting is its relatively low cost. Metrocable is lower because it is included in the integrated fare and transferring to the train and other public transportation modes does not need additional fare.

We also examined who uses Metrocable. The main users of Metrocable for commuting are female, low socio-economic class and live near a Metrocable station. Figure 10.3 shows head-of-household average monthly income and their travel modes for commuting. From this figure, we find that respondents with middle income use Metrocable at the highest ratio. The highest income group and the lowest income group rarely use Metrocable for commuting. One of the reasons is that around 40% of the highest-income group work at home or within walkable distance, and many of them likely have their own business. On the contrary, in the lowest-income group, most of the respondents are supposed to be informal workers who work at home or within walkable distance. We can find the similar tendency in terms of educational background.

Table 10.1 Commuting travel modes before and after Metrocable operation (only respondents who have same workplace)

	2003 (%)	*2016 (%)*
Bus	37.0	26.1
Train	7.6	22.7
Metrocable	—	26.1
BRT	—	2.5
BRT/train integrated bus	—	7.6
Car (driver)	1.7	4.2
Car (passenger)	48.7	21.0
Motorcycle	6.7	16.8
Only walking	5.0	4.2

Source: Matsuyuki et al. (2020b)

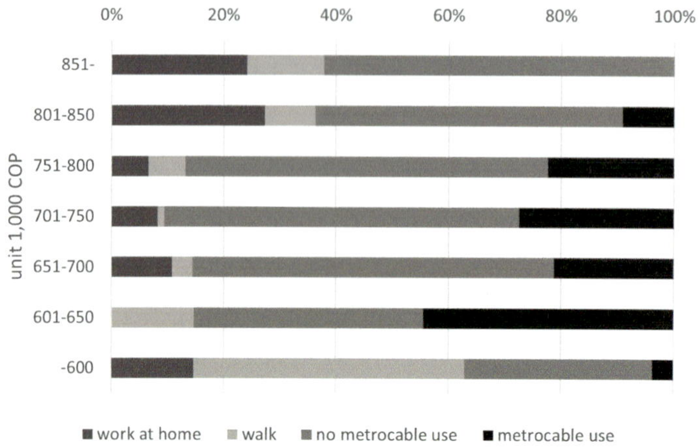

Figure 10.3 Monthly head-of-household income and Metrocable usage for commuting

Source: Matsuyuki et al. (2020b)

10.3.3.3 Job Opportunities

We examined whether Metrocable has led to vocational change and if that vocational change has increased their income. We found that 33.3% of respondents changed their jobs after Metrocable opened and almost 90% of the vocational change was a "spontaneous" vocational change, which aims for better job. Both those who changed job spontaneously and non-spontaneously increased their monthly household income from 2003 to 2016, from 222,000 COP (COP: Colombia Peso, 1 COP = 0.00035 USD) to 722,000 COP and from 323,000 to 748,000 COP, respectively. The income increase by spontaneous job change was higher than the income increases by non-spontaneous job change, and the difference is significant. Actually, we can say that spontaneous vocational change has led to income increase. We developed the model explaining the factor affecting the spontaneous vocational change and found that Metrocable usage promoted spontaneous vocational change, especially for lower and middle income and unskilled workers who live close to a Metrocable station, and the spontaneous vocational change has led to a higher income increase. Therefore, we can consider that Metrocable has indirectly increased monthly income.

10.3.3.4 Crime Reduction

We examined the effect of reduction of crimes in the area. In the questionnaire survey, we asked the frequency of criminal occurrence in their neighborhood in 2003 and 2016. To investigate the effect of people watching down from the cable car, we made a 100-m buffer from the Metrocable line and examined the difference of the average number of crimes reduced and crime reduction ratio between the inside and outside of the buffer area. The average number of crimes was reduced in both outside and inside of buffer: 13.76 cases to 1.66 case in a year and 19.04 cases to 1.29 cases in a year, respectively. There is no significant difference in the average number of crimes reduced and ratio of crime reduction between inside and outside of the buffer, and we could not find the effect of viewing from Metrocable on crime reduction.

10.3.3.5 Sense of Pride of Their Neighborhood

Next, the effect of Metrocable on improvement of sense of pride in their neighborhood was examined. We asked the level of pride that they feel about where they live on a five-point scale (1 = totally not proud of, 2 = not proud of, 3 = moderate, 4 = proud, 5 = very proud). The average score of users and non-users are 4.63 and 4.43, respectively, and this difference is significant. It suggests that the frequent use of Metrocable tends to enhance people's pride in their neighborhood.

10.3.4 Conclusion

The impact of Metrocable for commuting travel covers socially vulnerable groups, such as women and low socio-economic class, but we should take notice that the lowest income and group rarely use Metrocable for commuting and they do not receive direct benefit from Metrocable. The installment of Metrocable has led to vocational job change, and this vocational job change has led to higher income increase. Mainly, residents who have a child and earned lower to middle income and were non-skilled workers before Metrocable installation mainly changed their job spontaneously after Metrocable. Therefore, the past vulnerable group has earned the indirect benefit of Metrocable. Our research revealed that, on the one hand, Metrocable has no influence on crime reduction in the neighborhood area. On the other hand, usage of Metrocable increases passengers' senses of pride in their neighborhood. As Medellfn seemed to awaken from a dark period marked by economic stagnation, a lack of social cohesion and a sense of hopelessness arising from the violence that characterized the daily life of its inhabitants, especially the poorest, this increase in sense of pride is expected to contribute to further development of this area.

10.4 GENTRIFICATION BY INSTALMENT OF URBAN RAILWAY IN BANGKOK

Mihoko Matsuyuki

10.4.1 Introduction

Urban rail transit systems have been implemented globally to address transportation-related challenges, including air pollution, traffic congestion, and global warming. These systems often lead to increased property values and dense residential and commercial development around transit stations, known as TOD. While TOD has many positive effects, such as promoting transit ridership, it can also result in gentrification, displacing lower-income residents with wealthier individuals. This section explores the occurrence of gentrification in Bangkok's TOD areas and its impact on travel behavior among new and original residents.

10.4.2 Study Area and Methods

As described in Chapter 8, Bangkok's urban railway history began in late 1999 when the elevated railway system, called BTS, started its operation. Now it has three lines which have spurred significant condominium development around their stations. This section focused on 12 BTS stations areas along the BTS Silom and Sukhumvit lines (Silom line: Krung Thonburi, Wongwian Yai, Pho Nimit, Talat Phlu, Wutthakat, Bang Wa station, Sukhumvit line: On Nut, Bang Chak, Punnawithi, Udom Suk, Bang Na, Bearing station) in inner suburbs. A questionnaire survey was conducted with residents of newly developed condominiums and traditional shophouses within an 800-meter radius of the stations from October to December 2017.

The survey targeted two types of residences: condominiums, representing new residents, and shophouses, representing original residents. Condominiums are multi-unit residential buildings where owners share common areas, while shophouses are traditional Southeast Asian buildings with commercial activities on the ground floor and residential spaces above. A total of 230 respondents participated, including 164 from condominiums and 66 from shophouses. The questionnaire captured information on residents' socioeconomic characteristics, vehicle ownership, and travel behavior (Matsuyuki et al., 2020a).

Figure 10.4 Shophouses and Condominium along BTS lines

Source: (Photos taken by Mihoko Matsuyuki)

10.4.3 Socioeconomic Characteristics of Residents

Condominium residents are generally younger, with an average age of 33.8 years compared to 37.9 years for shophouse residents. Condominium residents generally have higher incomes and educational levels than shophouse residents. While the average monthly household income for condominium residents is around 58,000 THB, shophouse residents earn approximately 47,000 THB. Additionally, 93% of condominium residents hold a bachelor's degree, compared to 45% of shophouse residents. These disparities indicate that new residents moving into TOD areas are typically more affluent and better educated, fitting the profile of "gentrifiers."

Shophouse residents tend to have larger families, with an average household size of 3.6 compared to 1.9 for condominium residents. They also tend to have a higher proportion of business ownership, indicating a stronger presence of family-run enterprises. In terms of housing tenure, 55% of condominium residents own their homes, whereas only 26% of shophouse residents have ownership, with many renting their residences. This lack of ownership among shophouse residents makes them more vulnerable to displacement due to rising rents.

10.4.4 Travel Behavior – Commuting Travel Mode

The travel modes of condominium and shophouse residents differ notably. Condominium residents frequently use the BTS, with 68% of respondents using it as their primary mode of transportation to work. However, they also have a higher tendency to drive cars, with 18% commuting by car. By contrast, shophouse residents are more likely to use motorcycles and walk to work, with 19% and 12%, respectively. This divergence is attributed to the proximity of workplaces to residences for shophouse residents, often within walking or short motorcycle distances. Shophouse residents typically have their businesses nearby, enabling them to walk or use motorcycles for their commute, which supports a more sustainable travel behavior.

Table 10.2 Main travel modes of condominium and shophouse residents

	Condominium (%)	*Shophouse (%)*
Car	17.9	9.2
Motorcycle	1.9	18.5
Taxi	3.7	4.6
Motorcycle taxi	1.9	4.6
BTS (train)	68.5	41.5
Bus	0.0	1.5
Shuttle bus	3.1	6.2

Table 10.3 Car and motorcycle ownership

	Condominium (%)	*Shophouse (%)*
No car	17.0	35.4
1 car	75.5	52.1
More than 2 cars	7.4	12.5
No motorcycle	84.0	22.9
1 motorcycle	11.7	56.3
More than 2 motorcycles	4.3	20.8

10.4.5 Automobile Ownership

Table 10.3 shows car and motorcycle ownership; 83% of condominium residents own cars, a significantly higher percentage compared to the 65% of shophouse residents. Interestingly, many condominium residents reduced the number of cars they owned after moving to their current locations near BTS stations. The average number of cars owned by condominium residents decreased from 1.5 to 1.2 after moving to their current residences. This reduction can be attributed to the increased convenience and accessibility of public transit options, as well as the higher costs and limited availability of parking spaces in these high-density areas.

Conversely, motorcycle ownership is much higher among shophouse residents (77%) compared to condominium residents (16%). The average number of motorcycles owned by shophouse residents is 1.8, compared to 0.6 for condominium residents. These differences in vehicle ownership further influence the choice of travel modes, with shophouse residents being more inclined to use motorcycles and condominium residents being more likely to drive cars.

10.4.6 Impact of Workplace Proximity on Travel Behavior

One of the key findings is the influence of workplace proximity on travel behavior. Shophouse residents benefit from having their workplaces nearby, allowing for shorter commutes and the use of walking or motorcycles. By contrast, condominium residents, who may work further from their homes, are more reliant on the BTS and cars. The average travel time for shophouse residents who walk to work is just 8.1 minutes, compared to 19.0 minutes for condominium residents. Similarly, motorcycle commutes are shorter for shophouse residents, averaging 10.2 minutes versus 14.5 minutes for condominium residents.

This proximity effect is crucial in understanding the sustainability of travel behaviors. Shophouse residents' ability to live and work in close proximity reduces their reliance on private vehicles and promotes less-energy consumption modes of transportation, such as walking and motorcycles.

10.4.7 Conclusion

This study reveals that the new residents living in condominium in TOD area have higher incomes and educational levels compared to the long-term residents in shophouses, indicating that gentrification is occurring due to TOD development in Bangkok. While TOD aims to promote the use of public transportation and reduce car dependency, the findings show that condominium residents have indeed reduced their car ownership and increased their use of the BTS, suggesting that moving to TOD areas helps decrease car dependency and promotes rail use. However, as the original shophouse residents, who primarily relied on walking and motorcycles due to their proximity to their workplaces, are replaced by condominium residents, the overall car usage in the area may actually increase. This indicates that rail development and TOD do not simply reduce car usage; rather, the dependence on cars and the use of rail transit vary based on the characteristics of the residents living in the TOD area.

10.5 CONCLUSION

Mihoko Matsuyuki

This chapter first discussed the concepts of horizontal equity and vertical equity in urban transportation in Section 9.1. Horizontal equity emphasizes that people in similar conditions should equally share the benefits and burdens of transportation infrastructure and services. By contrast, vertical equity advocates for special consideration for socially disadvantaged groups, such as low-income individuals, the elderly, and persons with disabilities. Section 9.2 outlined Curitiba's efforts in designing its bus system and the impact of these initiatives. In Curitiba, routes targeting low-income areas were developed, offering measures such as fare exemptions for low-income residents and return bus tickets for public facility users. The introduction of the southern circular route also supported housing improvements and employment assistance in low-income areas. However, the increase in land prices along the structural axis led to gentrification, pushing low-income residents to peripheral areas.

Section 9.3 examined the impact of Medellín's Metrocable, which serves low-income hillside communities and integrates with rail services through unified fares. The survey found that the Metrocable system improved connectivity between low-income areas and the city center, significantly reducing commuting times. Some residents were able to change jobs and increase their incomes, while social effects such as enhanced community pride were also observed. However, the system's direct benefits did not extend to the poorest residents, revealing its limited impact on the most vulnerable groups.

Section 9.4 investigated gentrification and changes in travel behavior among new and long-term residents in areas along Bangkok's urban railway. Surveys showed that new residents, who lived in condominiums, were wealthier than long-term residents in shop houses. The replacement of shop houses with condominiums indicated gentrification. While condominium residents were more likely to use the railway, they also owned and used cars more frequently than shop house residents. By contrast, long-term residents tended to adopt a lifestyle close to work, using walking and motorcycles more often and exhibiting more sustainable travel behavior than new residents.

These findings indicate that in emerging economies, efforts are being made to provide affordable public transportation services to low-income areas as part of initiatives to improve both horizontal and vertical equity in transportation. These efforts can be seen as measures for poverty reduction. For instance, in Medellín, some residents improved their income by changing jobs facilitated by better access, demonstrating the effectiveness of transportation in poverty alleviation. However, not all low-income groups benefited equally. The poorest residents did not use the Metrocable for commuting, and the system's direct impacts on them were limited. Therefore, improving equity through transportation services must address disparities within social groups, not just between different income classes. Transportation strategies for poverty alleviation should carefully identify which subgroups within the low-income population benefit and which do not, and alternative measures should be implemented for those left behind.

Moreover, as shown in the cases of Curitiba and Bangkok, high-quality transportation services can lead to gentrification. In Bangkok, this phenomenon likely contributed to less sustainable travel behavior across station areas due to gentrification. According to the TOD concept proposed by Peter Calthorpe, it is recommended that various types of housing – differing in ownership, size, and price – be located within the same blocks or nearby. This approach promotes diverse TOD communities where people of different ages, ethnicities, and incomes can live and interact daily. While TOD is often interpreted as developing condominiums and shopping centers near stations, returning to Calthorpe's original TOD concept and considering housing affordability are desired.

Thus, transportation infrastructure can either promote or hinder equity in urban areas. Even within the same low-income group, some individuals benefit while others do not. When planning transportation infrastructure, it is critical to carefully evaluate who will benefit, who will not, and how these changes will affect the use of the transportation infrastructure and regional sustainability over the long term. Thoughtful consideration is necessary to ensure that transportation development aligns with broader equity and sustainability goals.

REFERENCES

Bullard, R. D. (2001). Environmental justice in the 21st century: Race still matters. *Phylon (1960-)*, *49*(3/4), 151–171. doi:10.2307/3132626

Hanson, S. (2010). Gender and mobility: New approaches for informing sustainability. *Gender, Place & Culture*, *17*(1), 5–23. doi:10.1080/09663690903498225

Litman, T. (2007). Evaluating transportation equity: Guidance for incorporating distributional impacts in transportation planning. *Social Research in Transport (SORT) Clearinghouse*, 8.

Lucas, K., Martens, K., With, F., Ciommo, D., & Dupont, A. (2019). *Measuring transport equity*. Elsevier. doi:10.1016/C2017-0-01787-1

Matsuyuki, M., Aizu, N., Nakamura, F., & Leeruttanawisut, K. (2020a). Impact of gentrification on travel behavior in transit-oriented development areas in Bangkok, Thailand. *Case Studies on Transport Policy*, *8*(4), 1341–1351. doi:10.1016/j.cstp.2020.09.010

Matsuyuki, M., Okami, S., Nakamura, F., & Sarmiento-Ordosgoitia, I. (2020b). Impact of aerial cable car in low-income area in Medellín, Colombia. *Transportation Research Procedia*, *48*, 3264–3282. doi:10.1016/j.trpro.2020.08.128

Peters, D. (2013). *Gender and sustainable urban mobility. Thematic study prepared for the 2013 UN-habitat global report on human settlements*. doi:10.13140/RG.2.1.4746.9287

Schweitzer, L., & Valenzuela, A. (2004). Environmental injustice and transportation: The claims and the evidence. *Journal of Planning Literature*, *18*(4), 383–398. doi:10.1177/0885412204262958

Vasconcellos, E. A. (2011). Equity evaluation of urban transport. In H. Dimitriou & R. Gakenheimer (Eds.), *Urban transport in the developing world: A handbook of policy and practice* (pp. 332–358). Edward Elgar Publishing.

Zuk, M., Bierbaum, A. H., Chapple, K., Gorska, K., & Loukaitou-Sideris, A. (2018). Gentrification, displacement, and the role of public investment. *Journal of Planning Literature*, *33*(1), 31–44. doi:10.1177/0885412217716439

Chapter 11

Transportation Systems and the Environment

11.1 HOW TRANSPORTATION AND ENVIRONMENTAL ISSUES ARE RELATED?

Fumihiko Nakamura

11.1.1 Definition of "Environment"

The relationship between environmental sector issues and transportation is generally complex. In the first place, it is important to note that the scope of the term "environment" varies widely from one person to another. The term "residential environment" is often used along with direct terms such as air quality, soil quality, water quality, vibration, and noise.

In this context, there are three different meanings, especially when referring to the atmospheric environment. Specifically, there are those in which local concentrations of atmospheric substances, such as nitrogen oxides and suspended particulate matter, are a problem; those in which dispersion over a wide area, such as sulfur dioxide, is a problem; and those in which the absolute amount is a problem, regardless of space, such as greenhouse gases that affect global warming.

11.1.2 Environmental Area in Buchanan Report

The term "residential environment" has been used in many countries, especially in terms of road traffic management (TM) in residential areas, since its definition in "traffic in towns" (the so-called "Buchanan Report") published in the United Kingdom in 1963 (Buchanan, 1963). The Buchanan Report classifies urban roads into four levels according to their functions: primary distributor, district distributor, local distributor, and access roads. For each roadway, the ratio of automobile traffic that has a use along the road (access traffic) and that does not have a use along the road (through traffic) is clearly divided. All auto traffic on the primary distributor is through traffic, while all auto traffic on the access road is access traffic. Each phase also gives an idea of the expected traffic volume, speed of traffic

 DOI: 10.1201/9781003467403-14

flow, and road geometry. The access road is not wide and has low traffic volumes and slow speeds. In addition, it denies the connection of roads that jump from one phase to another. For example, it does not allow access roads to connect directly to primary distributors. Under such a phasing structure, Buchanan names the area bounded by the primary distributors and/or district distributors as the Environmental Area and explains that no through traffic should enter the Environmental Area that has no business in the district. Buchanan compares the urban area to a house, with the primary distributor or district distributor as a corridor and the residential environmental area as a room.

Note that these theories are intended to show how urban residential areas should be configured. On the one hand, when applied to existing urban areas, they do not describe what traffic regulations or shape changes can be incorporated into existing streets, the procedures and methods for implementing them, or how to build consensus among the entities involved.

On the other hand, this concept is often utilized in the construction of new urban areas, such as in new town planning. The purpose of eliminating through traffic and reducing automobile traffic in residential areas is to reduce the risk of traffic accidents in areas where residents live, where there are many pedestrians, and where various activities are carried out on both sides of the road. Specifically, the goal is to eliminate automobile traffic that does not originate and terminate in the district, or so-called through traffic, and to reduce the volume and speed of other types of traffic as well. Thus, in primarily residential areas, the priority in daily life situations is to reduce the risk of accidents caused by automobile traffic, which is what is referred to here as improving the residential environment.

11.1.3 Strategies to Reduce Environmental Impacts by Vehicles

Automobile traffic on the road directly and indirectly affects each element of the environment other than the residential environment. In the case of gasoline-powered vehicles, the more they drive, the more emissions they produce and the more they affect the atmospheric environment, whether on the local, regional, or global scale described earlier. At signalized intersections on arterial roads, depending on the type and amount of large vehicles stopped, local nitrogen oxide concentrations can increase rapidly, and depending on the shape and topography of buildings along the road, winds may not flow easily, causing high concentrations to accumulate, which can have health effects on district residents. As for greenhouse gases, it is necessary to take into account that although they are not concentrations, they accumulate almost in proportion to the distance traveled, which has an impact on climate change and global warming. Although driving itself does not directly affect the soil environment or water quality, the manufacturing process of automobile vehicles, especially when considering the life cycle of the battery and other components, leads to various risks related to soil and water pollution. Depending on the pavement condition of the road, both noise and vibration can be generated by automobile traffic (World Bank, 2019).

Traffic demand is not necessarily met solely by automobile traffic on the road. Therefore, in solving the environmental problems described earlier, there are many ways to lower the environmental impact through technological innovations, but we will also take into account ways to reduce the amount of automobile traffic in the first place.

The four-stage estimation method was first applied in the United States and the United Kingdom in the 1960s. In Japan, it has been officially used since 1967 in traffic survey analysis in urban areas and future traffic demand estimation based on such analysis. The JICA uses the four-stage estimation method in urban area transportation planning in emerging countries, and has developed software for this purpose. The four-stage estimation method assumes a four-stage procedure for individual travel (trip) decision-making – starting, deciding the destination, deciding the means of transportation, and deciding the route – and estimates the amount of concentrated traffic, distributed traffic, transportation means sharing, and distributed traffic for each of these stages. The technical details of each stage are not discussed here, but understanding the four stages of mobility decision-making allows us to organize ways to reduce automobile traffic demand and lower environmental impacts by modifying the decision-making process.

Let's follow the previous four steps backward to give an example. First, let us take the example of route choice. If we can induce the demand for automobiles traveling on poorly paved roads to change their route to one with better pavement, noise and vibration can be suppressed. The next example is about the choice of means of transportation. In the case of personal use of automobiles, we can induce people to choose rail, water transportation, or carpooling. Those modes of transportation have less impact on the environment per trip length per person, thus resulting in less environmental impact. The next example is about destination choice. By guiding destination choice decisions to change plans to distant destinations to plans to nearby destinations, even automobile trips can reduce their trip length and thus reduce their environmental impact. The final example concerns the decision to depart. Trips will cease to occur if people stop going out in the first place, or if they stay home. Even if this is not such an extreme example, the logistics example is a realistic illustration. By reducing the frequency of traffic demand, for example, from several deliveries a day to twice a day from the warehouse to the store, the daily mileage of vehicles carrying goods can be reduced, thereby reducing the environmental impact.

This picture applies equally well to emerging countries as well as to developed countries. However, the composition of the problem is the same, as transportation options with low per capita emissions per distance, including the use of BRT and paratransit, have already been put to practical use. However, in general terms, among the current demand for car use, there are two groups: the captive group (those who do not recognize other modes of transportation as alternatives because they do not have or are not aware of any other options) and the choice group (those who have chosen cars but are also considering other modes of transportation and may change their choice if time, cost, and other conditions are met). The ratio of the "choice" group varies greatly depending on the income status of the country and the diffusion of transportation methods other than cars. It is well known that there are two aspects: one is that owning and using a car is a social status, and people use cars to protect and demonstrate it, and the other is that people use cars because of security concerns on the road and in public transportation vehicles. To overcome these issues, it is necessary to improve the security of public transportation in necessary situations and to improve the quality of its infrastructure and services, as a matter of course, but it is also necessary to devise comprehensive measures to change the social image of car-free travel. In reality, improving the security and infrastructure quality of all public transportation systems in emerging cities is a time-consuming and costly challenge, but it can be made more feasible by offering safe and high-quality public transportation options in some sections and by devising fare policies.

Note that various statistical indicators can be used to estimate the environmental impact of automobile traffic. However, vehicle performance often differs between 1) catalog values, 2) when new, and 3) when actually driven. Especially in emerging countries, vehicle emissions, noise, and vibration deviate from so-called theoretical values because there are many cases where vehicles are in a poor state of maintenance due to delays in technology diffusion and parts procurement, and because there are often considerable problems with the pavement condition of roads. Although it is time-consuming and expensive, it is necessary to conduct research and analysis based on actual measurements (whether on the road or at a vehicle inspection and maintenance facility) on vehicles actually used in the field.

As described earlier, the issues of transportation and the environment are involved in various ways.

11.2 VEHICLE ELECTRIFICATION AND ITS EFFICIENT MANAGEMENT

Fumihiko Nakamura

11.2.1 General Trend

The number of electrically powered vehicles is gradually increasing around the world.

This is due to the fact that many European countries have made major policy moves to promote electric vehicles; especially in China, electric vehicles are now mass produced and more affordable than ever before. Needless to say, however, electrification using electricity derived from renewable energy sources should be encouraged in terms of life cycle costs, and electrification will not contribute to decarbonization if a certain percentage of thermal power generation from coal or oil is used.

11.2.2 Electric Buses

Among so-called motor vehicles, the electrification of buses often attracts attention. The electrification of buses can be categorized into several types of power supply methods. In principle, electric buses are normally recharged once a day at a garage, but there are also examples of electric buses that are recharged multiple times a day. Some buses charge at the beginning and end of the day, while others charge at each stop along the way. A slightly older technology, the trolley bus, which runs on overhead wires installed on the road like a streetcar (strictly speaking, it is different from regular streetcars and LRTs because it has two parallel lines, one positive and one negative), is also a type of electric bus and, if classified, would be a bus with a constant power supply.

In the 1990s, cars in the form of continuous section cars, with diesel engines in addition to electric motors, were introduced in many parts of the world to run on sections without overhead wires. High-floor cars with diesel engines mounted in the first half of the section of a continuous section car, or in the second half of the section at the end of the train, were developed. More recently, relying on improved battery performance, trolleybuses have begun to emerge that run on battery power, especially in sections without overhead wires, without a diesel engine. If a trolley bus is equipped with a diesel engine, it is difficult to call it an electric bus, but if not, it is a type of electric bus and would be classified as a timely power supply type. Quick chargers are not large, but they are noisy, so consideration must be given to the living environment when installing them. In the case of technology derived from trolley buses, maintenance of overhead wires and the impact of the overhead wires and supporting steel poles on the road scape will be an issue. Battery life is a serious problem that has been pointed out for a long time, including in the case of non-trolley buses. Frequent battery replacement systems can be seen in tricycles and passenger cars, but it is not easy when the batteries are large.

In addition, efforts to simultaneously manage vehicle allocation (fleet management) and energy management have begun in many regions. The ability to manage vehicle status data in real time is expected to be utilized for efficient operations. In particular, real-time management of vehicle operations, incorporating the location and timing of recharging, is very important for both efficiency of overall energy consumption and efficiency of vehicle operations, and with recent advances in digital and computer technology, efficiency improvements can be expected. These topics are discussed, for example, when considering the terminal demand (from home to station or from station to home) at suburban stations of subways and other mainline railroads in large cities in emerging countries. There are not a few cases where relatively door-to-door service is required, depending on the distribution of residential density in urban areas within the station area. If development is more intensively guided, terminal bus routes may be feasible, but this may not be the case. In such cases, so-called on-demand or taxi-type service operations would be implemented with smaller vehicles, where fleet management in conjunction with energy management, as described earlier, could achieve lower carbon, higher efficiency, and improved service. In the field, there are issues such as contracting methods and money flow practices among drivers, vehicle owners, and service providers in the district, as well as how to coordinate competition among operators and between drivers, which cannot be solved simply by introducing technology on management. Various technological innovations have been made in the terminal logistics and mail delivery scene to link this fleet management and energy management.

The life cycle cost of electric vehicles is not cheap, and it will take more time to spread in regions where it is difficult to secure financial resources.

11.2.3 Electric Three-Wheelers

While much attention tends to focus on the electrification of four-wheeled vehicles, there have been several examples of electrification of vehicles other than four-wheeled vehicles, especially in emerging countries, since the 1990s. In Kathmandu, Nepal, where the city has a basin shape, an attempt to electrify cabs with three-wheeled vehicles began in the 1990s. Later, in Thailand and Bangladesh, electric-powered three-wheeled cabs (the names vary from region to region) have also been tried, although not on a large scale. In recent years, examples can also be found in Tanzania and other African countries. More and more three-wheeled cabs are being equipped with the same kind of technology as ride-hailing services in developed countries, whereby they are hailed via an application and the fare is settled on the cell phone, thus eliminating the negative image of trouble caused by negotiating fares. As mentioned earlier, there are many challenges in promoting management with new technologies, but it is also true that new technologies and services are gaining citizenship at the grassroots level.

Motorcycles are also beginning to be electrified. In Kunming, China, non-electric motorcycles are already banned in urban centers. Electric motorcycles are gradually increasing not only in China but also in Vietnam and Thailand.

In the electrification of motorcycles and tricycles described previously, the batteries are usually recharged sequentially, but there are also forms in which the batteries are replaced and the vehicle continues to run, in which case there is concern about soil contamination problems caused by discarded batteries. Nevertheless, by devising a mechanism for battery replacement and reuse of used batteries, electrification could be promoted more cheaply and efficiently for vehicles that are operated in a taxi-like fashion.

11.3 DEMAND MANAGEMENT TO REDUCE TRANSPORTATION SECTOR EMISSIONS

Yurie Toyama

11.3.1 Why Demand Management Is Important to Reduce Emissions?

Vehicle electrification is not the only possible measure to reduce emissions from the transportation sector. Controlling demand, reducing travel, discouraging the use of private cars through road pricing and other means, encouraging carpooling to reduce the number of cars, and increasing active travel such as bicycling and walking have also been considered.

These stories have much to learn from the United Kingdom and the United States, and we would like to give a brief history (Niitani and Harata, 2017).

11.3.2 History of Demand Management and Its Methods

In Europe, particularly in the United Kingdom, automobiles became popular early in the world, and traffic congestion, air pollution, and noise caused by the increased demand for automobile travel were recognized as problems early on. The purpose of TM is to reduce congestion and prevent traffic accidents, and does not include the pedestrian perspective. Buchanan (1963) redefined it as Environmental Traffic Management (ETM) from the perspective of environmental protection and residential safety. In the late 1970s, TM and ETM were further integrated and popularized by the concept of Comprehensive Traffic Management.

Meanwhile, in the United States, the concept of automobile traffic control was also evolving. Around the 1970s, as the environmental problems caused by road congestion became apparent, Transportation System Management (TSM) began to spread. The goal was to harmonize roadway operations, management, services, and policies to maximize the efficiency and productivity of the transportation system as a whole. This concept later evolved into TDM. Whereas previous management methods involved studying ways to increase transportation capacity in response to increased travel demand, the TDM approach was to control travel demand by controlling travel behavior.

In a narrow sense, TDM is the dispersal of demand over time and space, but in a broader sense, it includes land use, urban growth management, and location guidance that affect transportation demand. Specific TDM methods can be organized along five axes. The first is source adjustment. This includes adjusting the number of work days and using online tools to reduce travel. The second is changing the time of day. For example, staggering work hours reduce travel during peak periods. The third is changing routes. For example, road pricing, travel restrictions, and the provision of congestion information will be used to choose less congested routes. The fourth is to change modes. The fourth is to change the means of transportation: using public transportation instead of private cars, park and ride to avoid congestion in congested areas, encouraging bicycling and walking, developing infrastructure for this, revising fare systems, and so on. The fifth is the efficient use of automobiles. Since it is inefficient to travel in one car per person, this includes, for example, encouraging carpooling and other forms of vanpooling, and improving load factors through freight consolidation.

TDM is not just something that government encourages people to do. For example, in some cases in the United States, employers are required by regulation to provide TDM to their employees through commuter benefits and access to parking facilities. It is also conceivable that developers of large facilities could implement TDM in terms of parking and access to public transportation.

In Japan, since about 2005, various demonstrations have confirmed the effectiveness of Mobility Management (MM), which aims to reduce the environmental impact of transportation and to encourage people to change their attitudes and behavior from excessive use of cars to appropriate use of public transportation, bicycles, and so on (Fujii, 2007). MM is particularly unique in that it encourages transportation behavior change through communication, for example, when a travel feedback program is conducted to encourage people to reflect on their car use and consider future actions, while TDM is more about working with individuals. MM, like TDM, can also be designed for individual organizations such as workplaces.

11.3.3 Demand Management and beyond

An unexpected influence on these peak behavior changes was the pandemic in 2020. During the pandemic, people were encouraged to stay at home, the practice of online conferencing spread rapidly, and more people telecommuted. While this led to changes in housing and various other changes in residential and office areas, it also greatly reduced peak-hour rail congestion.

Some people switched to cycling and walking during the pandemic, and some governments encouraged people to do so, in order to avoid contact with others and to avoid confined spaces as much as possible. Paris is a typical example, where people were encouraged to spend 15 minutes in the city for various social activities, and some road spaces were modified and bike lanes built to encourage cycling.

In recent years, cycling, walking, and other non-motorized modes of transportation have been collectively termed "active travel" (Cook et al., 2022) and its health benefits have been actively promoted. In particular, the United Kingdom government has established Active Travel England, a government agency to actively promote Active Travel, and developed and distributed a toolkit to make it easier for local governments to promote and evaluate Active Travel.

11.4 COORDINATION OF SPATIAL PLANNING AND MOBILITY DESIGN FOR CARBON-NEUTRAL SOCIETY

Fumihiko Nakamura

11.4.1 Balance between Macro-Scale and Micro-Scale Approaches

Spatial planning here refers to the planning of land use on a macro scale and the planning of building floor use on a micro scale. Mobility design, a term coined by the authors, is a generic term for activities that consider the design of mobility, its usability, and management perspectives.

When it comes to decarbonization, the only goal tends to be to reduce automobile traffic, but if we think of this as a simple objective function, it would be enough if all people were to pull into their homes and somehow become self-sufficient. But if we go back to the 1990s' definition of sustainable development in the first place, both the local economy and society need to be sustained, balancing some sort of trade-off between them and environmental sustainability, and sharing the common but differentiated responsibilities of developed and emerging countries, since that time, will be a challenge.

11.4.2 Trade-Off among the Elements in the Concept of Sustainability

Based on this premise, spatial planning and mobility design are positioned to derive a well-balanced solution that takes trade-offs into consideration. Specifically, the spatial planning and mobility design should comprehensively consider various factors, such as land use, building floor use, and their interlocking effects, so that people can lead high-quality lives and economic activities even if they use cars infrequently, and over a short distance. There are various factors that vary from district to district. In general, they include pedestrian traffic flow, plazas and other placemaking areas, bicycle traffic flow, bicycle parking, location of shared bicycle stations, bus service content such as fares and schedules, location of bus stops, bus stop design, pedestrian traffic flow connecting bus stops to buildings and plazas, and traffic flow for handling vehicles. The above are all relevant. In addition to these, emerging countries also need to provide a variety of transportation services between buses and cabs. The ability to live a high quality of life and engage in economic activities without being forced to endure such things that are unlikely to last is also implied. In some cases, it may be necessary to change the values and behavioral norms that are implied in the term "behavior change". However, even without thinking too hard, things would change considerably if the design of the road network, the guidance of building use, and the guidance of building permits were not based solely on the use of automobiles. In the real estate business, car use is a prerequisite for buyers, but if the winds change little by little, it is hoped that individuals will change their behavior with relatively little difficulty.

11.4.3 Optimization and the Second-Best Solutions

However, in the application of short-term measures in a real city, it is impossible to find a so-called optimal solution, so the image is to seek the next best solution by coming to terms with the trade-off factors. Specifically, it is necessary to determine how to set the amount and content of travel demand, how to set priorities for each space or time, and how to set the processing capacity of each space. There are many stakeholders, and determining the next best solution will require coordination among all of them.

For example, in the city center, the main street should be operated as a pedestrian-only space during daytime hours, while encouraging the use of vehicles for cargo handling and venous logistics such as rubber processing during early morning and late night hours, and setting a boarding and unloading area for shared transportation services in the vicinity. Parking lots should be strategically located and priced, and parking fee revenues should be returned to users of other modes of transportation. This will reduce the number of visitors to the city center by private car, while maintaining the attractiveness of the city center itself, increasing the number of visitors, the time they spend in the city center, and the amount of money they spend. There will be a slight load on logistics delivery and on the car users of the captive population. The realization and continuous improvement of a package of measures tailored to such a scenario will contribute to decarbonization. For the management described here, data and communications for real-time management and sequential measure evaluation and monitoring based on accumulated data are useful.

These issues are not limited to urban areas in developed countries, but are rather useful in urban areas in emerging countries, where the use of cars and motorcycles is expected to further increase in the future. In addition, it is particularly necessary in the case of emerging country cities to ensure the maintenance of basic road and sidewalk pavements, and the security and ease of use of public transportation, including not only railroads and buses, but also various other things positioned between buses and cabs.

11.5 CONCLUSION

Fumihiko Nakamura

The relationship between transportation and environmental issues is extremely complex, and the scope of the term "environment" varies depending on the situation. The main physical elements are air quality, soil quality, water quality, noise, and vibration. In particular, air quality includes local pollution (e.g., nitrogen oxides, particulate matter), long-range dispersion (e.g., sulfur dioxide), and greenhouse gases that contribute to global warming (e.g., carbon dioxide).

The 1963 report "Traffic in Towns" (the Buchanan Report) provided an important perspective on the effects of traffic and street design in residential areas. This report divided the urban area into four categories: arterial, local, district, and access roads. It clearly defined the traffic mix for each. In particular, the concept of "residential environment" aims to exclude through traffic and ensure the safety and comfort of residents' living spaces.

Automobile traffic causes various environmental problems. Gasoline-powered cars increase exhaust emissions each time they are driven, which has a negative impact on the air environment. In particular, localized pollution can be a problem at intersections where a number of vehicles are stopped. In addition, the longer the distance traveled, the more greenhouse gases are emitted, causing climate change. To reduce these effects, it is necessary to reduce traffic volume and improve technology.

In addition, methods to reduce traffic volume have been proposed through the "four-step estimation method" of travel demand forecasting, which analyzes individual travel decisions. Specifically, changing the route, changing the mode, changing the destination, and even reducing the number of trips are all effective ways to reduce environmental impact.

The use of electric vehicles is becoming more widespread around the world, and their production and use are accelerating, particularly in Europe and China. The use of renewable energy is recommended for electric vehicles, but if they use electricity generated from fossil fuels, electrification will not necessarily contribute to decarbonization.

Electric buses can be recharged not only at the depot, but also at intermediate stops using quick-charging technology. Trolley buses are another form of electric bus, powered by overhead wires. The introduction of electric buses is accompanied by issues such as the development of charging infrastructure and the life of vehicle batteries, but fleet management and energy management are being linked to promote efficient operation.

In emerging countries, the electrification of not only four-wheeled vehicles, but also three-wheeled and two-wheeled vehicles is progressing. For example, electric three-wheeled vehicles are being introduced as taxis in Nepal, Thailand, Bangladesh, and some parts of Africa. These vehicles operate efficiently through a battery swapping system and are expected to contribute to a low-carbon society.

Spatial planning, on the one hand, is about deciding how to use land and buildings. Mobility design, on the other hand, is about designing with ease of movement and ease of use of transportation in mind. The goal of a decarbonized society is not only to reduce the frequency of car use, but also to improve the quality of life and economic activity. This requires striking a balance between reducing traffic and achieving a sustainable society.

Achieving a sustainable society requires a balance between the environment, the economy, and society. For example, simply reducing the frequency of car use is not enough; we need to find ways to reduce car use while maintaining a high quality of life. By changing policies related to land use and building design, we can increase the options for non-car modes of transportation and improve the sustainability of cities.

In actual urban planning, it is difficult to find the ideal optimal solution, and it is more realistic to consider different trade-offs and come up with the next best solution. For example, it is necessary to come up with designs and operational methods that reduce traffic demand while maintaining travel quality. In addition, improving the convenience of public transportation, buses, and walking will contribute to the realization of a low-carbon society.

REFERENCES

Buchanan, C. (1963). *Traffic in towns: A study of the long term problems of traffic in urban areas*. London: Her Majesty's Stationery Office.

Cook, S., Stevenson, L., Aldred, R., Kendall, M., & Cohen, T. (2022). More than walking and cycling: What is 'active travel'? *Transport Policy*, *126*, 151–161.

Fujii, S. (2007). Mobility management as an integrated transportation policy: A continuous management that integrates soft and hard measures. *Transport Policy Studies' Review*, *10*(1), 2–10 (in Japanese). doi:10.24639/tpsr.TPSR_10R_01

Niitani, Y., & Harata, N. (2017). (Ed.). *Urban transportation planning* (3rd ed.). Gihodo Publishing Co (in Japanese).

World Bank. (2019). *Leaders in urban transport planning: Workshop report*. Author. Retrieved from https://www.worldbank.org/en/news/feature/2019/07/10/leaders-in-urban-transport-planning-workshop-report

Conclusion

Toward Sustainable and Inclusive Urban Transportation Systems in Emerging Countries

Mihoko Matsuyuki and Fumihiko Nakamura

This book aims to investigates urban transportation systems in cities of emerging countries, focusing on the characteristics, challenges, and underlying contexts of various issues in each transportation mode. It also explores the adaptation of new technologies and the modernization of transportation management. Additionally, it explores interrelationships between transportation urban elements. From the perspective of emerging countries, the book seeks to present practical approaches for building sustainable and inclusive transportation systems.

As described in this book, urban transportation systems in emerging countries have unique characteristics and challenges arising from rapid urbanization and economic growth, which differ from those in developed countries. These characteristics can be summarized as follows:

First, emerging countries are experiencing rapid urbanization and an explosive increase in transportation demand, yet they face significant shortages in road, public transportation, and logistics infrastructure. This situation has led to severe traffic congestion and decreased efficiency, significantly affecting urban life. At the same time, diverse transportation modes – including walking, cycling, buses, railways, and paratransit – coexist, providing flexibility and responsiveness. However, traffic congestion, safety issues, and a lack of regulation remain major challenges. Additionally, the shared use of road space by freight transportation and urban transportation has exacerbated congestion and reduced efficiency, highlighting the absence of dedicated freight infrastructure.

Paratransit plays a complementary role in public transportation, offering flexibility, yet it faces safety and reliability challenges. Effective management, regulatory improvements, and the use of digital technologies to enhance service efficiency will be crucial moving forward. While systems like BRT and urban railways are being introduced, their operational coverage remains limited, and coordination with other transportation modes is insufficient. This lack of integration reduces user convenience and increases the burden of mobility.

Moreover, the misalignment between land use and transportation planning has exacerbated urban sprawl and forced low-income groups to relocate to urban peripheries, thereby widening social inequalities. Public transportation systems often fail to sufficiently address the needs of vulnerable populations, such as low-income individuals, the elderly, and people with disabilities. Although there are successful examples, such as Medellín's Metrocable, challenges remain in adapting these solutions to other regions.

From an environmental perspective, the impacts of transportation-related emissions and noise are severe, compounded by delayed regulations and standards. While the adoption of electric vehicles and renewable energy is progressing, addressing policy and cost challenges is essential for broader implementation. Additionally, short-term economic growth is often prioritized, which may lead to the neglect of long-term sustainability and holistic transportation planning.

In emerging countries, while challenges abound, there are also notable strengths. Among these, the adoption of digital technologies stands out as a distinctive advantage. Technologies such as mobile payments and transportation apps are increasingly being utilized, though regulatory barriers and costs continue to hinder the integration of comprehensive transportation information systems. Despite these challenges, emerging countries demonstrate exceptional flexibility and speed in adopting and disseminating new technologies and services, which can be considered a competitive edge over developed countries.

Developed countries can learn a great deal from emerging countries, particularly regarding the rapid adoption and dissemination of new technologies and services in urban transportation. The effects and challenges of these early implementations offer valuable insights. Additionally, the coexistence of paratransit, which can be regarded as a form of micro-mobility, and private vehicles – a phenomenon commonly observed in emerging countries – provides lessons for developed countries attempting to introduce micro-mobilities.

When formulating urban transportation plans for emerging countries, it is crucial to fully understand these distinct characteristics, which differ significantly from those of developed countries. Furthermore, the diverse contexts of individual countries and cities within emerging markets must be recognized. Tailoring plans to the specific conditions and characteristics of each city is vital for ensuring effective and sustainable transportation solutions.

To address the challenges faced by transportation systems in emerging countries, a shift is needed from policies prioritizing short-term economic growth to those emphasizing long-term urban transportation planning. The integration of land-use planning with transportation planning is crucial, particularly in designing cities that are livable and accessible for low-income populations. Furthermore, rather than simply imposing stricter regulations on informal transportation, efforts should focus on improving safety and service quality while fostering collaboration with formal transportation systems.

The inherent flexibility and adaptability of emerging countries represent a significant strength in developing transportation systems. Leveraging this advantage, low-cost and efficient operational management systems utilizing mobile apps and digital payments can be implemented. Additionally, expanding transportation services tailored to the needs of local residents and harmonizing informal sectors with transportation development can help sustain urban vitality while driving progress.

International technical and financial support is vital in overcoming the challenges faced by emerging countries. For instance, assistance from developed countries and international organizations can facilitate the development of transportation infrastructure and the adoption of environmentally sustainable technologies, contributing to both immediate improvements and long-term solutions.

In addressing the challenges of urban transportation, it is essential to recognize the complexity of these issues especially in emerging countries. For example, as noted in Chapter 4, while paratransits have seen improvements in service quality through app-based platforms, this trend has also introduced new forms of informality. Given this complexity, it is crucial to evaluate the effects of proposed solutions from multiple perspectives.

We should also recognize that the major cities in emerging countries are undergoing rapid and profound transformations, and much of what is written in this book will continue to evolve. One area expected to change significantly is information and communication technology (ICT). Advances in ICT are likely to dramatically enhance the efficiency of managing and utilizing transportation systems. Improvements in computational speed, data storage capacity, and communication technology will enable real-time processing and forecasting of transportation data, with the widespread adoption of artificial intelligence (AI) further boosting efficiency. Furthermore, these technological advancements are anticipated to extend beyond the direct improvement of transportation services, influencing a wide range of areas such as transportation education, governance, and human resource development. At the same time, as the global movement of people, goods, and information accelerates, urban transportation systems may face new risks and challenges. Flexible and adaptive responses to these emerging issues will be essential to ensure the resilience and sustainability of future transportation systems.

In this rapidly changing society, the approach to future urban transportation planning is also expected to undergo significant transformation. Traditionally, transportation planning has relied on methods such as the four-step estimation process to predict future transportation demand and subsequently develop infrastructure and services, such as roads and public transportation, to meet those projections. However, in an era characterized by rapid and unpredictable changes, this approach has reached its limitations. Moving forward, there will be a growing need for planning methods that adopt a backward perspective, focusing on desired outcomes, and for adaptive strategies that respond flexibly to emerging changes and uncertainties.

While adaptability to such changes is essential, the fundamental principles of transportation planning – "safety" and "smoothness" – remain unchanged. Moreover, principles increasingly emphasized today, such as "sustainability," "equity," and "inclusivity," are equally vital and enduring in urban transportation planning. Even in emerging countries, it is crucial to base urban transportation strategies on these principles while continuously adapting to ever-changing environments.

Index

3 Es (Engineering, Education, Enforcement) 39, 62
5G 129
15-minute city 167; *see also* x-minute city
20-minute city 167; *see also* x-minute city

access restriction 58, 59
access road 208, 209, 218
active transportation 6, 13, 15, 32–33
active travel 214, 216, 219
ADAS (Advanced Driver Assistance Systems) 130
ADB (Asian Development Bank) 43
affordability 3, 26, 190
AGT (Auto Guided Transit) 106, 109
AI (Artificial Intelligence) 223
air pollution 1, 2, 4, 8, 56, 72, 104, 137, 170, 190, 193, 201, 214
air quality 9, 57, 69, 208, 218
Almere 89
Amsterdam 167
angkot 81
animal-powered vehicle 71
annual license tax 27
ARL (Airport Rail Link) 170, 172
asthma 193
Australia 167; *see also* Melbourne
automated driving 127, 128, 130, 131, 134
automotive industry 8, 127, 134, 141
autorickshaw 71

Bangkok 1, 5, 8–9, 15–16, 42, 66, 87–88, 111–112, 114, 136, 142, 147, 152, 155, 169–175, 184, 186, 194, 201–202, 205–206
Bangkok Metro Company Limited 170
Bangladesh 213; *see also* Dhaka
Barcelona 13
barrier-free 190–192
Beijing 27–28, 30, 112, 117
bicycle parking 32–33, 158, 217
bicycle sharing 5, 6, 26–28, 30–33, 132
Big Three 27
bike lanes 168
BMR (Bangkok Metropolitan Region) 171–175, 185–186
BMTA (Bangkok Mass Transit Authority) 66
Bogotá 4, 5, 7, 81, 89, 90, 92–97, 100–102, 105, 158
Bosphorus Strait Tunnel 105
BOT (Build-Operate-Transfer) 118
bottleneck 119
branch line 81, 92, 94, 96–99, 148, 159–160, 182, 195
Brazil 15, 22, 89–90, 92, 98, 114, 151, 160; *see also* Curitiba; Porto Alegre; São Paulo
BRT (Bus Rapid Transit) 7, 23, 31, 68, 80–82, 86, 88–90, 92, 94, 96–98, 101–102, 105–106, 109–110, 121–122, 134, 137–138, 146–147, 157–158, 169, 177, 185, 199, 211, 221
BTO (Build-Transfer-Operate) 118
BTS (Bangkok Transit System) 111, 170–172, 202–205; *see also* Skytrain
BTSC (Bangkok Mass Transit System Public Company) 170
Buchanan report 208, 218
bus stop 17, 23–24, 91, 99–100, 102, 147, 152, 160, 168, 184, 195–196, 217

CACC (Cooperative Adaptive Cruise Control) 130
Calthorpe, Peter 166, 168, 206
Cambodia 66, 72, 132, 136; *see also* Phnom Penh
car ownership 2, 25, 27, 40, 43, 107, 121, 170, 178–180, 185–186, 205
car pooling 139
cardiovascular diseases 193
caregiving work 192
CASE (Connected, Autonomous, Shared, and Electric) 8, 127–128, 130, 134, 141
CDM (Clean Development Mechanism) 158–159
Cheonggyecheon 97, 100
chicken-and-egg problem 129
China 1, 5–6, 15, 25–33, 36, 38, 90, 117, 212, 214, 219; *see also* Beijing; Shanghai; Wuhan
Ciclovía 4, 5, 92
CNG (Compressed Natural Gas) 72
Colombia 4, 7, 90, 92–93, 102, 158, 196, 200; *see also* Bogotá; Medellín
commission system 87
community activity 14, 32
commuter rail 108, 110, 114–115
commuter railway 105
commuter train 7
commuting subsidy 27
compact city 166, 167
complementation 137
compliance 40, 49–51, 82
comprehensive traffic management 214
comprehensive transportation information systems 222
concession 43
conductor 73, 91, 139
conflict point 47
connectivity 63
contract 76, 80–82, 94–95, 116, 131, 156, 194, 213
conventional railway 108, 113–116
convoy 89, 102
COP3 158
corruption 3, 55, 57
cost structure 76
coup 3
courier services 135
crime reduction 200–201
crosswalks 48, 50, 168
Cultural Friday 93
Curitiba 6, 8, 15, 22–25, 33, 89–92, 95–98, 100–102, 105, 151, 154, 160, 169, 182–186, 194–195, 205–206

Da Nang 15, 18, 20–22, 33
decacorn companies 135
dedicated lane 63, 98
Delhi 112, 117
demand structure 81
Dhaka 112, 154, 160
DiDi Bike 28, 30–31
digital information 146, 147
digital payments 223
digital technologies 221, 222
digitizing data 8
disadvantaged groups 9, 191, 194, 205
discrimination 2, 191
dispatcher 74, 77
disposable income 27
distance-based fare 120
district distributor 208–209
driver licensing systems 4
driving license 39, 49, 63, 88, 137, 192
drone delivery 61
drop-offs and pick-ups 2, 45
DSRC (Dedicated Short-Range Communications) 129
DX (digital transformation) 139, 147

economic disparity 190
economic inequality 2, 193
EDSA (Epifanio de los Santos Avenue) 59
egress 44
Ekachai 87, 88
electric and hybrid cars 40
electric bus 9, 212–213, 219
electric three-wheeler 9, 213–214
electrification 9, 107, 128, 133–134, 212–214, 219
elevated railway system 202
elevated structure 108–109, 157
emission regulation 1
emissions control 4
e-money 135
enforcement 169
England 216
entrepreneurship 56
entry regulation 78
environmental area 208–209
environmental pollution 169

environmental quality 14, 167
equity 8, 42, 137, 163, 167, 186, 189–191, 193–195, 197, 199, 201, 203, 205–206, 224
ERP (Electronic Road Pricing) 129
e-scooters 132
ETC2.0 129
ETM (Environmental Traffic Management) 214
EV (Electric Vehicles) 57, 104, 133–134, 212–213, 219, 222
externality 194

fare collector 73, 75, 145
fare coordination 8
fare systems 7, 8, 119, 141, 159, 215
fare-leakage 154
feeder 69, 81, 82, 110–111, 115, 119–120, 122, 147, 169
fill and go operation 76
financial management 116, 157–158
financial strategy 8
financing 30, 81–82, 113, 117, 158
first/last mile mode 138
fixed price 69, 76, 141
flat fare 119–120, 154, 184
flower Street 24; *see also* Ohana Dori
food delivery 126, 141
food stalls 15
for-hire transit 65, 78
formal transportation 67, 80, 223
formalization 7, 79–81, 83
Fortune Public Bicycle 28
four-step estimation process 223
France 167; *see also* Paris
franchised bus 66
freight infrastructure 221
freight transportation 7, 55–56, 113, 221
fuel 1, 25, 36, 40–41, 61, 75, 77–78, 219
fuel tax 25, 40

gated community 171
gender 190
gentrification 9, 189, 191–192, 194, 201–202, 205–206
Gett 132
GHG (Green House Gas) emissions 56, 134, 137
global warming 201, 209–210
Gojek 132, 135, 137–138
Google map 142, 146–147
governance 3, 66, 142, 223
governmental capacity 3
GPS (Global Positioning System) 30, 61, 99, 145
Grab 83, 132–133, 135, 137–138
green time ratio 47
growth management 166–167, 185, 215
GTFS (General Transit Feed Specification) 146
Guangzhou Asian Games 27

The Hague 167
Hangzhou Public Bicycle System 28–30
Hanoi 8, 16, 37–38, 62, 88, 147, 169, 175–178, 181, 185–186
Hanoi Metro 177
health care 2
health impact 1
health risks 193
heavy rail 109–110
Hello Bike 28, 30–31
helmet 32, 83, 132–133, 137
Helsinki 156
hierarchical network 119
historical discrimination 191
Ho Chi Minh City 88, 112, 147
holistic transportation planning 222
horizontal equity 8
human-powered vehicle 71

IATSS (International Association of Traffic and Safety Sciences) 39, 62
IC card 81, 100, 119, 147, 153, 154
ICT (Information and Communication Technology) 5, 7, 8, 33, 56, 96, 100–101, 125–127, 139, 141, 223
IDA (Initiative for the Development of Africa) 60
Incentive 22, 79, 82, 95, 130, 133, 136
inclusivity 190, 224
India 1, 72, 105, 114, 117; *see also* Delhi; Pune
Indonesia 1, 5, 38, 71, 96, 114, 117, 127, 132–136, 140, 142; *see also* Jakarta
infill development 167
informal public transportation 145, 199
informal sector 3, 223
informal transportation 3, 66–67, 80, 145, 223

informality 7, 65, 67–68, 70, 79, 81, 83, 223
information provision 8, 144–147
initial cost 65, 77, 109, 111, 116–118, 121, 131, 133
institutional transportation 67
insurance 28, 83, 130, 137
Int. V 19
Interchange 8, 59, 120, 144, 148, 150
interchange management 8, 148
intercity rail 109–110, 141
international aid 89
International technical and financial support 223
Internet Plus 30
Intersection 13, 39, 46–48, 53–54, 78, 92, 95, 209, 218
IPPUC (Institute de Pesquisa Planejamento Urbano de Curitiba) 22–23, 194
Istanbul 89, 105, 114, 119
IT-based bicycle sharing system 28
ITDP (Institute for Transportation and Development Policy) 82, 90, 96, 103

Jakarta 42, 81–82, 96, 112, 114, 117, 134, 138, 142, 147
JakLingko 82, 147
Japan 2, 3, 39, 63, 86, 90, 92, 115, 116, 125, 129, 131, 140, 152, 156, 167, 182, 210, 215; *see also* Tokyo
Japan's Metropolitan Area Development Plans 167
jaywalking 48, 64
jeepney 70, 72, 82, 134, 147
JICA (Japan International Cooperation Agency) 92, 210

Katmandu 213
Kenya 60, 64, 73, 82, 114; *see also* Nairobi
Kenya Railways 114
Keolis 131
Khon Kaen 15, 17–22, 33
Kingdom of bicycles 25, 27
Kuala Lumpur 114, 117
Kyoto Protocol 158

labor cost 77, 100, 117
labor regulation 78–79
labor shortage 130, 139
land access 41
land use 3–4, 6, 8, 23, 42, 57, 61, 116, 120, 142, 165–188, 215–217, 219, 222
land use patterns 3, 6, 169
land-use regulations 4
leapfrog 5, 60, 125, 140
Lee, Myung-bak 97
legal framework 4, 57, 61
Level 4 autonomy 141
license holder 74–75
license plate regulation 55
licensed bus 66
local cargo handling center 59
local distributor 208
logistic estates 59
logistics 6, 7, 37, 55, 57, 210, 213, 217–218, 221
low-income group 8, 68, 79, 88, 101, 154–156, 160, 191, 193–194, 206, 222
low-income individual 2, 222
low-income population 3, 68, 222
low-income resident 9, 68, 91, 96, 154, 184, 186, 196–198, 205
LRT (Light Rail Transit) 89, 109, 111–112, 117, 121–122, 147, 184–185, 212
LVC (Land Value Capture) 168
Lyft 132

MaaS (Mobility as a Service) 8, 127, 138–141, 146, 155–156, 160–161
Malayan Railways 114
Malaysia 117, 140; *see also* Kuala Lumpur
Manila 42, 58–60, 111–112, 114, 147
market-responsive 65, 67
Marmaray 105, 114
mass transit 1, 2, 7, 66, 68–69, 80–82, 89, 92, 97, 101–102, 104, 106, 110, 120, 132, 137, 138, 144, 170, 177
master plan 8, 22, 24, 91, 110, 112–114, 148–150, 159, 169, 175, 177
matatu 70–71, 73, 82
maximum loading capacity 50
maximum speed limit 51
Medellín 9, 158, 194, 196–198, 205–206, 222
medium transit systems 137–138
megacity 1, 59, 104, 112, 117, 168
Meituan Bike 28, 31

Melbourne 167
Metro Manila 58
Metrocable 9, 196, 201, 205–206, 222
Mexico 1, 5; *see also* Mexico City
Mexico City 5
micro mobility 222
microtrans 147
middle class 2, 171, 175
military regime 3
millimeter waves 129
minibus 72, 81, 92, 95, 147, 158
minority 190–193
mitigation 61, 191
MM (Mobility Management) 215
Mobike 30
mobile apps 223
mobile payment 222
mobile phone 5, 100, 140–141
mobility design 9, 216–217, 219
mobility hub 152, 160
modal share 27, 41, 56, 177, 180
MRT (Mass Rapid Transit) 80, 108–111, 114, 117, 121, 147, 170, 172
MUD (Mixed Use Development) 166–168, 185

Nairobi 42, 70–71, 114–115
necessary activity 19, 21–22, 33
Nepal 213, 219; *see also* Katmandu
Netherlands 89, 167; *see also* Almere; Amsterdam; The Hague; Randstad Region; Rotterdam; Utrecht
New York 5
NMT (Non-Motorized Transportation) 44–45
no overtaking regulation 52
noise 9, 14, 56, 190, 208, 210–211, 214, 218, 222
non-motorized mode 41, 119, 216
notification 79, 82, 145

ODA (Official Development Assistance) 3, 43, 117–118
ODD (Operational Design Domain) 130
Ofo 30
Ohana Dori 24; *see also* Flower Street
one-stop 138
online food delivery 135
online grocery shopping 135
on-street parking 38, 47–48, 52–53, 100
open data 146–147
open space 21, 107
optional activity 19, 21–22
Oregon 167, 184

Pakistan 71
pandemic 105, 216
paratransit 2, 3, 5, 7, 65–83, 115, 119, 125, 132, 135–138, 140–141, 145, 147, 152, 211, 221–223
Paris 95, 128, 167, 216
park-and-ride 120, 139
parking 22–24, 32–33, 38, 42, 47–48, 52–53, 62, 100, 104–105, 120, 144, 156, 158, 177–178, 180–181, 185, 204, 215, 217
parking regulation 52–53
passenger capacity 50–51
passenger transportation 8, 144
pedestrian areas 6, 33
pedestrian space 6, 13–15, 32–33, 100, 107, 119
pedestrian street 13–15
pedestrianization 24
people with disabilities 154, 160, 190–191, 222
petrol tax 92–93
Philippines 1, 38, 58, 60, 70, 72, 82, 130, 136, 140; *see also* Manila; Metro Manila
Phnom Penh 66
pick-up, drop-off 45
planned economy 27
platform apps 67
platformer 72, 83
political instability 3
pollutant 69, 107
polycentric urban development 166, 185
polycentric urban structure 8, 175, 181, 185
Portland 167, 184
Porto Alegre 89, 102
potential user 144
poverty 2, 194–195, 198, 206
PPP (Public Private Partnerships) 3, 43, 117–118
preferential fare 154, 160
pricing 43, 119–120, 139, 214, 200
primary distributor 208–209
private transportation 68, 80, 83–84, 107, 120–121, 144–145, 156
procurement cost 116

profitability 43, 80, 111
property tax 177
property value 184, 191, 201
public awareness 32–33
public plaza 13
public space 6, 14, 24, 93, 100, 160
Public Utility Vehicle Modernization Program 82
PUI (Proyecto Urbano Integral) 196
punctuality 87, 89, 95, 106–107, 115, 175
Pune 1

Randstad Region 167
real-time bus operation management 95
real-time information 61, 139
real-time location 61, 145–147
recreational activity 14
Reform and opening up policy 26
regime changes 3
registration 3, 37, 57, 60, 71, 78–79, 82, 145
regulation 30, 33, 48, 51, 57, 77, 79, 82–83, 215, 221
regulatory regime 138
relocation 189
remorque 136
renewable energy 212, 219, 222
RH (Ride-Hailing) 5, 7, 8, 127, 132, 135–138, 213
ribbon development 169
ride-sharing 8, 127, 141, 155
right-of-way 157
ring road 42, 176, 178–179
road congestion 26, 52, 63, 77, 91, 115, 215
road density 46
road development 7, 40–44, 104, 117, 193
road infrastructure 41, 42, 48, 51, 56, 63, 66, 131
road maintenance 1, 43, 52
road network 1, 2, 42, 46, 55, 63, 98, 148, 217
road pricing 214–215
road safety 39, 42, 51, 56, 63–64, 70, 89, 119, 132
road space 13, 32, 38–39, 42, 45, 62, 89, 98, 121, 216, 221
road surface condition 47
rolling stock 108–112
Rotterdam 167
route finder 146–147
route-fixed transit 65, 71, 76, 81, 83
RT (Rapid Transit) 88, 108
Runcorn 89

SACCO 82
Sakay 147
São Paulo 5
satellite cities 178
SDGs (Sustainable Development Goals) 32–33
Seattle 89
security management 95
segregation 192
self-financing 81
self-regulation 75, 79
semi-formal 74, 81, 82
sense of community 14
sense of pride 201
Seoul 7, 96–101, 110, 112
Seoul Vision 2006 97
sewerage systems 14
Shanghai 28, 30, 42, 112
Shanghai World Expo 27
shared bicycle 5, 25–26, 29–31, 217
shared mobility 67, 69, 71–72, 78–79, 83–84
sharing economy model 27
shophouse 17, 171, 202–205
sidewalk 14, 16–18, 21–22, 33, 38, 45, 48–49, 53, 58–59, 168, 192, 218
signage 113, 146–147
Si lo lek 87, 88
Singapore 129, 135
Skytrain 170; *see also* BTS
smart city 8, 126–127
smart fare media 154
smart payment method 8
smartphone 5, 8, 30, 100, 125–129, 138–140, 147, 155–157, 160
social equity 21, 42, 167
social exclusion 191, 194
social experiment 4, 24
social inequality 222
social interaction 14
social justice 190
social status 2, 27, 45, 51, 62, 175, 211
society 4.0 141
South Korea 96–97, 125; *see also* Seoul
special tax 8, 158
speeding 32–33, 51, 61
sprawl 2–4, 8, 166–167, 169–171, 174–175, 185–186, 222

stakeholder 3–4, 7, 61, 217
static information 144–145
station plaza 3, 169
stationary activity 18–22, 33
stay at home 216
street food stall 216
street tree 14
street vendor 16–17, 33, 48
streetcar 35, 89, 109, 212
structural axis 23–24, 33, 91, 96, 182–184, 195–196
subsidy 27, 79, 82
substitution 38, 137
suburban railway 105–108, 114
suburbanization 170, 174, 179
subway 88–89, 100, 105, 108–109, 117, 140–141, 148, 170, 194, 213
Summer Streets 5
Superblock 13

Taipei 89
Taniguchi, Cassio 195
target system 75–76
tax revenue 40, 43, 93, 152, 158
tax system 40, 43
taxation policies 57
TDM (Transportation Demand Management) 9, 215
telematics 61
Terminal Appointment Booking System 60
territoriality 69
Thailand 1, 15–18, 33, 38, 63, 70–72, 87, 125–127, 130, 133, 135, 140, 170, 184, 213–214, 219; Khon Kaen 15, 17, 18–21–22, 33; *see also* Bangkok
three-wheeled motorcycle 71
three-wheeler 9, 66, 70–72, 213
ticketing 127, 139–141, 153
timeframe 19–22
time reliability 104, 106, 119–120
timetable 87, 95, 97, 100, 125–126, 139, 144–145, 147, 149, 159
TM (Traffic Management) 1, 39, 48–49, 53, 55, 57, 59, 63, 132, 208
TNCs (Transportation Network Companies) 132
TOD (Transit-Oriented Development) 8, 23, 96, 107–108, 166–169, 182–186, 202–203, 205–206, 224
Tokyo 104–106, 110, 114
toll road 42–43
TomTom Traffic Index 1
tourism MaaS 140
tout 74–76
townhouse 171
trade-off 9, 216–217, 219
traffic accident 38–39, 48–52, 56, 61–62, 104, 126, 134, 141, 189, 190, 193, 209, 214
traffic capacity 46–48, 54, 105
traffic congestion 1, 2, 6–7, 42, 44–49, 53, 55–56, 59, 61, 63–64, 97, 102, 104, 106–107, 134, 137, 169–170, 189, 201, 214, 221
traffic control 18, 48, 54–55, 215
traffic demand 44–48, 52–55, 60, 63, 113, 210
traffic demand management 48, 55
traffic engineering 17, 39, 42
traffic jam 44, 51–52, 136, 175
traffic norm 48
traffic police officer 54
traffic regulation 48, 51, 54–55, 209
traffic rule 48–49, 54, 78
traffic safety 6, 39, 50, 52, 54, 58, 63
traffic segmentation 49
traffic signal 47, 50, 54
traffic violation penalty 4, 49
train fare 175
tram 89, 108–109, 197–198
TransJakarta 81–82, 96, 134
Transdev 131
transit alliance 81, 118, 146
transit mall 24, 93
TransMilenio 89, 92–96, 102
transportation apps 156
transportation cost 190–191
transportation education 223
transportation hub 19, 76, 147, 160
transportation information systems 8, 222
transportation node 93, 107, 150–152, 166
transportation regulation 7
travel speed 46, 106–107, 109, 120–121, 182
travel time 2, 106, 119, 121, 144, 190, 197–199, 204
tricycle 66, 70–71, 147, 213, 214
trip planner 138
trunk line 81, 92, 95, 97–98, 119, 148, 195

trunk road 42, 151
TSM (Transportation System Management) 215
tuktuk 70, 72, 126, 135–136

Uber 132, 135
underground railway 106
underground structure 110–111, 116, 169
the United Arab Emirates 41
the United Kingdom 89, 208, 210, 214, 216; *see also* Runcorn
the United States 2, 131, 167, 182, 193, 210, 214–215; *see also* New York; Oregon; Portland; Seattle
universal design 150, 190–191
urban expressway 42–43
urban growth boundary 167
urban planning master plan 169, 175, 177
URBS (Urbanização de Curitiba Sociedade Anônima) 22, 23, 194, 195
used car 41, 72, 206
utility 82
Utrecht 167

V2X (vehicle to everything) 128–129
vehicle age restriction 4
vehicle axle load 57
vehicle electrification 9, 212, 214
vehicle inspection 211
vehicle licensing 57
vehicle ownership 4, 45, 179, 192, 202, 204
vehicle size 45, 58, 60, 67
vertical equity 9, 189–190, 205–206
Vientiane 38, 66
Vietnam 15, 18, 33, 37–38, 61, 71, 83, 88, 125, 127, 136, 175, 177–178, 181, 214; *see also* Da Nang; Hanoi; Ho Chi Minh City
violation 4, 49, 55
vulnerable group 189, 194, 201, 205
vulnerable social group 193

wagon 72
walkability 15
welfare 94, 154, 160
well-being 13, 14
wheelchair 14
women 16, 38, 132, 137, 190–192, 201
working condition 92, 137
working environment 65, 70, 83, 137
World Bank 43, 89, 117, 158
Wuhan 26

Xiamen bicycle skyway 31
x-minute city 167

Yandex 132
Yangon 66, 88, 115

For Product Safety Concerns and Information please contact our EU representative GPSR@taylorandfrancis.com Taylor & Francis Verlag GmbH, Kaufingerstraße 24, 80331 München, Germany